# Most By Nature

Matchstick Literary
1-888-306-8885
orders@matchliterary.com

# Most By Nature

Few are
intelligent

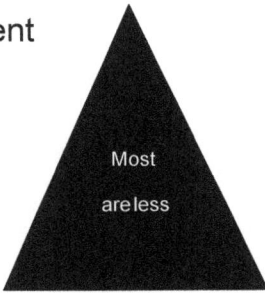

Few are less
intelligent

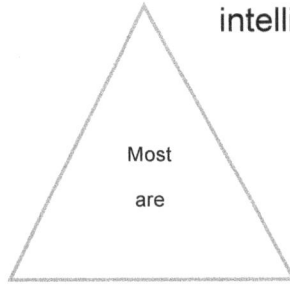

Most
are less

Most
are

black

nonblacks

## E. Asamoah-Yaw

# DEDICATION

For my children and children's children.

# ACKNOWLEDGEMENT

THERE ARE A few people who have made this work possible. Without them, this book would have been saved forever in my computer. I am really grateful to Mr. George Antwi, popularly known as Uncle George, and his wife, Sister Akos, both of New York City, for their great support in bringing this controversy to the general public. They deserve no blame with regard to the book's content. If there is anything blameworthy, I alone should be held accountable.

Justice Debrah, retired high court judge, did the initial proofreading with numerous important critiques. From my heart, I thank him for his inspirational support.

I would also like to thank Nana Osie of Abuakwa, Mr. Stephen Akonnor Boafo and Mr.

Evans Baidoo of Kumasi Girls and Ms. Mary Kraah former Headmistress also of Kumasi Girls Senior Secondary School; Mr. Kwaku Esamoah, my good, smart friend who always comes with unexpected suggestions and critical viewpoints on every critical discussions; local FM radio stations, especially Fox, Metro, Adom, and several others who offered me the chance the amplify my unique idea of black people's nature and how to correct them. These need to be mentioned with sincere appreciation of their intellectual discussions about our own type of humans—the black man.

All the medical doctors I interacted with for the past four years, too many to mention, helped me a lot to understand the intricate and real meaning of human beings as a very complex organic matter. Science has gone very deep into understanding nature, but there is still a lot we need to know than what we already know. As most would admit, black people have problems; solutions are still pending. But the major problem is the diagnosis of what exactly is wrong and the root cause of it. My doctor friends encouraged me to keep asking questions about my nature as a black person.

The Xlibris publishing staff actually transformed the inner feelings of my nature into what you are reading today as a book. Without them, this information would have been locked up in my engine of thought and into the grave.

I am thankful to the dozens of acquaintances who inspired me to say loud and clear what I think of my nature and the best way I can contribute to change the mind-set of my brothers and sisters for their own good. Thank you for everything.

# CONTENTS

# INTRODUCTION

AT THE TIME of writing this note, my advanced age of seventy-four should discourage me from sharing this perplexing odd feeling that I have always had from the age of about seven or nine years old. I was always the smallest and the youngest among all my friends and the one who tirelessly seek answers for weird questions.

We were six friends playing football while barefoot in front of my grandparents' house in the village of Abodom. There was a sudden unusual humming sound in the sky, far beyond the birds' flight range, and there was something that looked like a huge bird flying across the sky. All of us stopped playing and started staring and talking about that object. It was our first-ever encounter with seeing anything so big flying so high above the tallest tree around the village! The thing was about a million times bigger than a vulture, the biggest bird we had ever seen in the village. It was beyond our imagination as young innocent and deep-black African kids who had never traveled beyond one mile outside the village. Our entire world was the village.

At dinnertime around five in the house compound, I was scared of asking my grandparents about that huge monster I saw in the sky. Even knowing they would shout me down as they always did for asking too many questions, I asked Grandpa anyway if there was any bird bigger than the vulture and flew up higher in the sky.

Everybody laughed, and Grandpa retorted, "Haven't I told you not to talk when you are eating, that your mother will die when you eat and talk at the same time?"

I shut up instantly.

My real question about that specific object lived with me unanswered until the age of fourteen, when one of my early schoolteachers gave a talk about something called an airplane, a man-made machine that flew like birds and had people and heavy objects inside it. The classroom scene instantly reminded me of my childhood encounter with the sight of that sky object. The teacher added that they were manufactured in Europe by white people. The teacher's lessons opened another curious door in my mind. Many questions were forming up in my organ of thought.

On reflection of my childhood days, I can remember that almost everything around me then and even now at age seventy-four is foreign made. Many of them, especially the quality ones, are of European or American origin.

Now looking closely at me as a black person, there is no question about my nature that I am a human being because I possess all obvious features just like those of white people. But I kept wondering why more white people were and still are able to create objects that can imitate humans and other natural beings' acts but more of my kind of humans cannot. What gives white people the capacity or ideas to manufacture things with raw materials that most black people cannot? Black people cannot even imitate many of the things nonblack people have invented. We blacks simply do not possess the natural chemistry to initiate originality.

According to *The Black Book*, authored by Middleton Harris and three others, there is evidence of dozens of patented black inventors in US history. Hence, it sounds inappropriate for me to condemn black people as unimaginative. There are a few among every kind of people who are naturally different or gifted. My worried observation is that too many blacks in most black communities, unlike most nonblack people in most of their communities, do not progress and indeed appear to be comfortable with very simple, ordinary life. Most blacks don't exhibit a need to change.

Necessity for innovation of mind-set is mostly absent in most black people's conscience.

Is my nature as a black person different, or is it my environment which has molded me to be what I am? If we are all humans, why do some humans have the ability to create, to invent, to modify, or to originate something completely unimaginable by others? Among humans, my type is completely docile, unprogressive, and very contented with mediocrity.

We are nurtured to believe that all humans are equal because all species show identical natural traits when compared with other animals. We have also learnt that there are variations among humankind because of nature and nurture. There are certainly many obvious differences in appearances and body chemistry to the extent that each of us possess individual natural signature with our fingerprints. We are therefore doubtlessly completely different from one another as human beings yet very similar at the same time.

Indeed, some academics believe racial intellectual levels are equal or must be equal. I see this as too simplistic.

It is this natural trait of human *similarity*, *equality*, and *differences* which have prompted me to share my thoughts with curious minds like you, the reader, to see if this phenomenon is human nature or human nurture.

For the past five years, my devotion to search for facts regarding black people's lack of progress or matching parity with nonblacks never ceased. Starting from Black African sources, nothing much was discovered apart from our ancient histories with trivial, insignificant relics. I have come to the conclusion that there is a fundamental natural difference between *most* blacks and *most* nonblacks.

I have also attempted to find solutions or bring blacks into equal status with nonblacks through modern scientific means, whereby the identified major biological difference can be magnified and corrected medically, clinically, surgically, or through some other scientific means. I know this may sound preposterous, but I don't think so.

You are advised to read the entire discourse before pronouncing instant judgment. I am definitely aware of general public reaction, especially from fellow blacks and white apologists, with regard to my rejection of conventional wisdom of the environment as being responsible for *most* black people's ineptitude.

Honestly, the biggest question I intend to answer in this discourse is whether the best of the best amongst humans can ever be a black person.

Please keep reading.

I am sincerely honored for your time spent with me.

EA-Y

# CHAPTER 1

# Why Are We
# What We Are?

ALL OF US are products of *heredity* and *environment*. Firstly, the genetic materials we inherit from our mothers and fathers and, secondly, the habitat in which we are conceived and born and nurtured—these two elements operate and collaborate to produce, to be born, and to be maintained. The two together design us and also control us. Our genes *and* our habitat unquestionably determine who we are, why we are what we are, how we are, and what we are. Our personalities, values, features, temperament, successes and failures, intelligence and lack of it, plus every trait are functions of *nature and nurture*, according to conventional wisdom. How much each contributes to our being is a big debate.

We inherit our parents' blood chemistry during fusion of our father's sperm and mother's egg.

For approximately thirty-eight weeks in our mothers' body, millions of organic processes and changes take place to form us to become human beings.

After birth, the environment and our immature internal organs take over our growth and mold us. The organ of thought (*brain*), in particular, is believed to shape our culture, and also that culture and lifetime experiences equally shape the nature of the brain.

It is not as yet determined by thinkers if the two products (inherited traits and environment) form us in equal proportions—or that the ratio of the two forces (nature and nurture) is 50 percent on each side. The big question is, can the ratio be 51 and 49, 60 and 40, 70 and 30, or possibly, 99 percent and 1 percent. In short and without doubt, we are products of both simultaneously. Fundamentally though, our origin is doubtlessly our parents' blood chemistry.

*Firstly* and *factually*, all humans are *formed* through sexual intercourse of a female and a male. Alternatively, fusion of male sperm and female egg in the mother's body is the beginning of all humans. Immediately after fertilization of the two gametes, the sex organ of the seeded woman begins creating or designing a human being—*all by herself.* Practically, we are formed and kept in our mother's body, nourished, and delivered to this world by our mothers *only.*

Practically also, for thirty-eight weeks approximately, no one or nothing else performs that action apart from the mother or carrier of the fetus. Moments after delivery from our mother's womb to the world, the environments in which we are born or raised become a significant *additional* molding factor. We have become a real human matter who occupies space and time, although immature. Our blood cells adapt to surrounding circumstances to preserve and to protect the physical body.

Before delivery, therefore, the life of the fetus inside her body depends on our mother's body chemistry, and has to adapt to her prevailing habitat conditions.

As long as we eat good food, drink good water, breathe good air, and live in a hygienic or clean environment, we will grow as old as possible. Our internal organs or our physical body, while growing from infancy to maturity, adapt to local norms and surroundings in order to maintain life and behave normally as all others. Unquestionably, therefore, all humans and other organic matters are necessarily biologically and environmentally *equal;* we naturally thrive on air, water, light, earth, and fire/heat.

All organisms have a lifespan, unlike inorganic matters. Men and women are equal biologically, different physiologically, and similar anatomically by nature.

And because of these basic natural dispositions, it shall be argued here strongly that performance among and within species in their attempts to maintain life is primarily highly controlled by the natural quality of their genetic content *rather than* the acquired environment sustenance.

In my summary, I will extract three major puzzling elements in organic matters, especially in human beings, for critical analysis. The three are the following: *similarities*, *equalities*, and *differences* (simply abbreviated in this discourse as *SED*. Henceforth, this acronym (SED) will be used to represent all the above three primary inquiry topics.

The word *natural* shall be defined here as anything that is not man-made. *Man-made* is also defined here as anything created, manufactured, produced, formed, or constructed by organisms, especially humans.

In this definition, therefore, human beings are naturally human-made organic matters.

# CHAPTER 2

# SED: Similarities, Equalities, Differences

1.  Why are there so many differences and similarities in everything natural?

2.  Why are some things more similar and others so completely different although they fall into the same natural family group?

3.  Why do some objects of the same genus possess common similarities by appearances yet are different in reality when dissected?

4.  Why are some organisms 100 percent equal and so completely identical visually

or physically yet completely different by content and performance?

5. Should the existing SED in organic matters be questioned?

6. Are these SED the actual elements which glue all organisms together to form what we call the universe or that which separate and enrich our individualities?

## SED in Plants

THERE ARE BILLIONS of plant species. Each one is genetically different, yet when separated into groups of similar types, there exist common traits among each group. As an example, pine trees, tomato plants, grasses, potato plants, peanut plants, apple trees, plantain plants, orange trees, and cedar trees are different kinds of plants, yet they are classified as plants because they contain a unique natural property which makes plant species different from other organisms.

All trees are *trees*, but some trees grow from the smallest fragile basic stem structure at nursery level to taller bigger trees. Some are naturally short and small and will remain so throughout their entire life span. Others are naturally huge and massive with a lifespan exceeding 100 years. All plants grow leaves. Some creep on the ground, while others are strings and climb on other plants. Some produce their younger ones from roots. Others bear fruits with seeds for reproduction. There are indeed several thousand differences, equalities, and similarities.

Oranges, lemons, limes, tangerines, and grapefruits are naturally similar. They produce similar fruits with differences in sizes and tastes. In one sense, they are different, but in general, they belong to the citrus genus. Oranges may taste sweet. Lemon and lime may taste bitter and sour. Tangerine may taste smelly and sweet. Grapefruit may taste bitter, sweet, and sour. Each tree has its unique genetic composition.

Naturally, we cannot plant an orange tree and expect it to yield tangerine fruits (although it is now scientifically possible through genetic engineering to make an orange tree bear orange fruits and any of the citrus fruits). Each one is *different*, *similar*, and *equal* simultaneously. Why?

What is the basic constitution which makes each plant what it is? It is very tempting though to simply say that they are so because of their nature, but each species's environment can alter some genetic bearings through the survival adaptation process. Plants need to survive by adapting to their environment, or else they'll die. Nature and nurture together make all these possible.

# SED in Space-Based Animals

There are millions of bird species, but birds of the same feather always flock together. Those belonging to the same genus possess unique body chemistry which differentiates them from other similar types. Apart from the group's equal organic identity, each bird also carries its own genetic signature, which specifically belongs to it alone. This phenomenon is akin to all organic matters.

Birds of different species don't commune together. They are not naturally bound by blood type. All eagles, for instance, possess identical cells which make them different from hawks, but the two species look alike. A bird will always remain an airborne creature, a fish will always be a fish swimming in the same water habitat, and so will land surface animals naturally thrive within the habitat they belong. A monkey will remain a monkey although its body chemistry is about 5 percent different from humans. The 5 percent difference will keep the monkey in its nature always.

And a human (*Homo sapiens*) will forever remain the most rational creature when compared to all life because of the natural genetic difference.

A crucial point here is, must we accept the prevailing *natural* similarities, equalities, and differences as inalienable and unalterable natural facts of life, or should we ignore these and rather take them as merely *environmental* and therefore socially amendable traits? And if so, must we also accept that in all species of the same kind, there exists natural quality variation? There are indeed *genetic* inferiority, superiority, rationality, and efficiency variations measured by survival techniques and progress possessed by each organic group member. Again, with regard to performances (i.e. making birds' nests), all birds are naturally equal, yet under common tests, some birds' nests look more sophisticated than others. Some birds naturally fly faster and act cleverer than others even among birds of the same genus.

It will be unreasonable to reject as insignificant the natural quality difference in performances. It is indisputable that *similarities*, *equalities*, and *differences* in birds and other organic matter are primarily natural.

For instance, a test of intelligence among all airborne creatures selected from all continents shows that the parrot is the cleverest.

A second test to identify the smartest parrot also showed that the African continent–based parrot species excelled better than all other parrots. This conclusion demonstrates that the African parrot possesses natural qualities or blood chemistry which causes that bird to perform better than all parrots and all birds in other continents. The unique environment in which the African parrot thrives could surely be a significant contributory driving force behind the bird's intellectual power. But we cannot ignore the bird's body chemistry, its neurons, and the cells responsible for capturing, processing, and calculating information that ultimately define its nature as being the crucial cause of its intellectual supremacy.

This natural dynamic energy source should be seen as the primary motivating element responsible for the bird's brain performance. It is its nature first to be what it is. Its habitat is not the first or its major propeller. The environmental circumstances undoubtedly contribute or *add* some qualities to its performance. The African parrot's internal organs count first.

Birds and other flying animals have equal natural characteristics. They possess unique genes which make them what they are—for instance, vertebral skeletal structures with a head, a tail, wings, and beaks. Their central nervous systems are located inside the head. Their nerves pass through the backbone from the skull to the various parts of the body.

Thought mechanism and information processing and actions are equal with humans and other vertebral creatures.

Birds have the instinct to interact and communicate by sound and vision. They procreate through sexual interaction between males and females as humans do. A major exception is that after sexual interaction, an egg is laid and incubated *outside* the female body. Birds and human instincts are naturally different yet equal in the sense that survival, preservation of the self, and desire to create an offspring are similar.

Nurturing the newly hatched birds is also not different from how humans attend to their newly born babies. Teaching and learning, feeding and sheltering, security and sound moral practices toward the maturity of infants and attainment of independence are similar with humans.

Their natural body constitution is the basic element or engine which causes birds to act. The environment only contributes. Environment doesn't initiate.

Besides all these *similarities, equalities,* and *differences,* humans and birds are equal when observed in their broken-down natural groupings. Birds of the same feathers always naturally flock together in the same way most humans of the same species or race also generally and naturally commune together. Naturally by performance, some birds are creative and intelligent in the same way as some human beings are.

Some flying creatures are naturally very good at making or imitating meaningful sounds and collecting materials from their surroundings to build nests and dwell in them. All birds look for food and life sustenance needs within their habitats. As each bird possesses a unique gene for construction of shelter or nest, so do humans in that there are some who have the unique gene for constructing or creating shelters or houses.

Every bird species make its unique kind of nest to befit its nature. For instance, some birds are capable of making only very simple birds' nests, while some have instinct or the capability to make sophisticated nests. It is observed that some birds make nests only for protection of their eggs and newly hatched birds. These types of birds do not make enclosed nests for shelter or protection against predators.

These types of birds evacuate their nests when hatched birds mature. These acts of both birds and humans are natural. If you change a bird's environment to a completely different one, it will try as much as possible to construct the same nest it is capable of making given the gene it possesses. A bird transported from one continent to another continent or from the wild to domestic cage will have to adapt to its new habitat to survive, but its gene that influences the way it builds its nests can make the bird search for similar building materials to weave the same nest which its naturally inherited gene is capable of making. Birds in captivity will build nests with the same structure as those built by birds of the same group in the wild, although the facade may vary somehow because of climatic and material differences.

In general, birds do not progress in the sense of the self/environmental innovations because they don't appear to possess progressive genes as humans do in general. Birds and humans share the same organic structures; we are both vertebrates. Yes, birds and humans are naturally similar, equal and different at the same time.

E. ASAMOAH-YAW

Yes, again, there is a research which shows that among all birds, parrots are the most intelligent birds. *If* we can accept intelligence variations among birds as natural, we can also accept the same natural intelligence variations among human racial groups. Some humans are definitely naturally much smarter than other humans. In general, there is no doubt that most nonblacks are more intelligent than most black people.

## SED in Aquatic World (Fish)

Similarly, *all aquatic creatures* naturally differ in so many ways, yet there is a unique vital natural property or genetic content which makes them live or survive only in waters rather than on the land surface or in the air. Aquatic creatures are naturally *equal* because all of them live and swim in water. Yes, they swim. They do not fly nor walk or crawl. Fishes are different species.

Some dwell in salty waters only, and others in freshwater only.

Basically, each one is a fish or an aquatic animal, yet they have several millions of *natural differences* by their appearances, shapes, sizes, characters, performances, and inherited body chemistry or genetics. Even among those who are classified as belonging to the same group, some do have their unique genetic or natural differences. All fishes, big and small, are alike yet different.

The SED in aquatic creatures should equally be solely natural, but of course, a drastic change in their environment for a long period can completely alter their being and behavior. They are what they are because of their natural dispositions and their prevailing favorable environmental conditions.

Ocean fishes thrive in salty waters, while lake and river fishes thrive in freshwater. Because of their natural differences, their habitats are not the same. Unlike the space and land creatures, fishes' blood temperature is cold compared to other beings, and that is their nature. There are preys and predators that are colorful and not colorful, stupid and clever, big and small because of their nature, not necessarily their habitats.

A test of brainpower among all fish species shows that the dolphin is the cleverest in the aquatic world. An intensive test within the dolphin group also shows that the pink dolphin's brainpower is one of a kind. It is far more efficient than all remaining dolphins. It is the unique nature or biochemistry of the pink dolphin that makes it a select species with innate capabilities for absorbing and processing information in its world better than all other dolphins. Again, there is no doubt that the adaptation of the pink dolphin to its unique, cold Antarctic environment contributes to its unique display of excellence.

One other interesting discovery in my studies of dolphins' nature is a revelation that because of dolphins' unique intelligence, some cultures in India, Chile, Costa Rica, and Hungary call dolphins as nonhuman persons.

It is important, however, to recognize that *contribution* means an added factor, part of a whole, or at the most, less than 50 percent significant par. Dolphin's habitat helps or contributes to its basic biochemistry to exhibit efficiency. Thus, in the aquatic world, pink dolphins' intelligence surpasses all fishes mainly because of its innate makeup or its nature.

Firstly and most importantly, it is the biological nature of pink dolphins which causes it to be more intelligent in the aquatic world. The same analogy should apply to human species variations—that it is firstly and most importantly the biological nature of *most nonblack* people which cause them to be more efficient and *intelligent than most black people.* Black people are different from other humans because we possess more melanin in our body than all other humans. Apart from natural physical characteristics, melanin is Negroes' most obvious natural difference.

I suspect very strongly that our low efficiency level and our low intellectual capability are caused by the abundance of this natural element in our body chemistry because this is our most obvious scientific natural difference that's indisputable.

Among all fishes, the exclusive quality of dolphins' neurons causes it to capture, to process, and to act or perform better in the aquatic world.

## SED in Land Surface Creatures

*Land surface animals,* from ants to elephants to humans, are all labeled as living beings who ordinarily do survive only on land surface. A clear distinction exists between us and them. We humans simply call all organic beings *animals*, especially because of their natural physical characteristics, such as looks, features, and behaviors. We *Homo sapiens* love to separate ourselves from other organic matters although we share common natural elements: air, water, fire/heat, and earth. Humans possess unique natural capabilities to utilize these basic elements or matter for self-preservation, procreation, and domination.

We successfully subdue nonhumans with our intelligence. Yes, this is our essential difference. We thrive on land surface in the same way as birds do in the sky and fishes do in water.

Humans are *skillful* by nature in fabricating, manipulating, and using natural resources to improve health and living conditions.

We even sometimes successfully alter natural order to befit human needs. We make and utilize invisible air waves band to communicate and to reproduce sounds and pictures, both still and in motion, thousands of miles apart to improve quality of life. We progress. Human rationality surpasses all organic beings. Indeed, we don't have to be scientists to say that humans are *far more prudent* than all species on planet earth. It will also be safe to say that despite human beings' natural homogeneity, humans are naturally heterogeneous by every life criterion.

We are equal. We are different. We are also similar. Some humans are more equal than others. When humans are grouped into ethnicities (i.e. cultures, races, tribes, nations, and continents), it will be absurd not to admit that natural differences *between* groups and *within* groups is a real, indisputable observed natural fact among humans.

There are certainly many other natural differences. There are many natural similarities and at least 85 percent of all humans are genetically equal.

Some human groups or clans are naturally efficient than others. Some are naturally inefficient and dull. Some are naturally proficient and others clumsy. Some are naturally smarter than others. Some are naturally shorter; others are taller. Some are naturally pretty; others are ugly. Some are nasty by their nature; others are pleasant. Facial features differ significantly by nature. Some are less intelligent by nature, both within and between racial groups. Some are more inquisitive and resourceful than others. The same natural traits that are found in animal groups should also be seen to be found in humans as well.

Admittedly, environmental factors do contribute to these negative and positive natural differences, but environmentally also, it should be seen as something that influences a permanent organic part inside the body that is certainly not the base root cause of proficiency, efficiency, and smartness. There exist biochemical differences between races.

The origin is firstly biochemistry and secondly geochemistry. Foundation of the human body as an organic unit is *firstly* our nature and *secondly*, after birth, our nurture. From our mother's womb, we arrive at this world completely naive and without *mind* nor articulate information stored in the brain, although we are completely formed with all vital organs during the thirty-eight incubation weeks and ready to continue growth to fully matured person. *Quality* differences among humans are fundamentally, doubtlessly, and naturally inherited from our natural parents' *DNA first*.

A test of all human species or varieties must necessarily show the existence of brain quality differences in the same way we see differences in birds and in fishes. There exist *natural* performance quality variations among human species and all organic beings. We are *unequal* naturally—inwardly and outwardly.

Although this observation is from a layman's perspective, scientifically, all lives are equal because all life thrives in the same egg-shaped cosmic world called the planet earth.

None can survive without the uniqueness of earth with its composition of hydrogen, oxygen, carbon, nitrogen, and others such as phosphate. Planet earth is what it is because of these known natural life-sustaining ingredients or assets. Every life has a limited lifespan because the proportionate volume of each survival ingredient for all life is limited and fixed at every given moment. Each organic matter naturally creates offspring and perishes at long last to enable new ones to succeed and thrive. None will exist indefinitely.

Our existence is ultimately determined by our individual abilities to acknowledge and to rationally utilize the natural resources available.

It is a fact that *there are intrinsic natural similarities, equalities, and differences among all creatures, including humans, in all habitats.*

This is so among all organic groups and within each group members. This natural order or patterns of matter is a well-known phenomenon since time immemorial and therefore not new. It will be irrational indeed for us to deny the true nature of *SED* as a natural phenomenon.

Now let us isolate humans from all the other animals by physical characteristics, such as appearances, first and later by other means. Humans, monkeys, and apes are more *equal* and *similar* by natural anatomic characteristics yet very *different* genealogically. Human beings' major *obvious* difference is our skin and physical appearances. For instance, apart from skin complexion, *most* black people have thick lips; heavy, flat large noses; dark-brown or black eyes; wooly, spongy, and curled black hair. Black people's structures and internal organs are equal with the rest of humankind. All of us possess vertebral skeleton with heads jointly connected to our spine. Our central nervous system is the brain which serves as the organ of thought.

All visible organic beings' heads contain hearing organs, visual organs, vocal organs, taste organs, breathing organs, and sensory nerves.

*If* for natural reasons some parts of our engine of thought, the brain, have a cell whose duty is to distort information reception, processing, retention, and transmission, then the performance of the body that carries that cell will naturally suffer brain deficiencies.

And this is the core of my exploration. Does such a cell exist in all humans or some humans or not? Surely this question should be asked and answered by a scientist such as neurologist, a brain surgeon, a geneticist, a biochemist, a pathologist, and many such experts (definitely not an economist or a social scientist such as myself). I am so passionate about the nature of *most* black people (our historical comparative intellectual weaknesses, insensitivities, decadence, and perpetual dependence on nonblacks for every human need) such that hopefully there will be a final solution.

# CHAPTER 3

# We Are Our Nature First and Our Environment Second

IF ALL VISIBLE and invisible living organisms were capable of answering this question (why are we what we are?), I wonder what their answers would be. Are we what we are by nature alone or our environment alone or something else or by some of each or by none of the above? Many thinkers in the past for many centuries have pondered on these puzzling questions, but none have a universal consensus or answer. There appears to be a universal agreement that SED has a lot do with nurture than with nature—that our habitat counts more than anything else. Another big question is, to what extent is this generalization true?

As part of the animal group, human beings, with our rational superiority over all lives, attempt to answer the question why we are what we are with numerous logics. Most ancient major reason is the notion that besides this physical optical world, there is another invisible world existing concurrently with the visible one we see in our everyday life. That supposed other world is the spiritual realm, a world which exists only in an individual's mind-set.

Some accept that the spiritual world is the source of everything, that we are formed and managed by invisible forces, or a supreme unidentifiable entity.

For several centuries, everything that is inexplicable with human faculty is necessarily divine based, according to religious faith. This observation teaches that apart from nature and environment, there is a third causal force, the invisible, divine source force; to some people, it is a god.

Primitive ancestors of humans explained nature and the causes and effects of all worldly matter and occurrences with the idea of a supernatural or a spiritual being (omnipotent) as the entity responsible for not only forming the universe and its contents but also of perpetually conducting all affairs of life infinitely and invisibly.

Thousands of years ago, our ancestors' reasons for our being what we are were based on their knowledge or understanding of nature as experienced within their exposed environment at the times they lived.

The acquired wisdom was inherited and perpetuated by succeeding generations as facts. In the ancient times, for example, these human

E. ASAMOAH-YAW

habitats of a few families settled in shelters made with mud and wood as family homes. Each settlement was viewed as "the world" or "the nation."

The nature of their world then was practically a reflection of their locality. Each group of human settlement represented a unique cultural world of its own and spiritually controlled by the clan's god.

As human population increased, human settlements expanded. The concept of "the world" gradually changed to embrace new settlers, seldom with refined perception. Extended families required new settlements and expansion of social needs and responsibilities.

Some generations later, the concept of hamlets, villages, towns, cities, and countries gradually changed. Thus, the word *world*, which currently represents a global unit embracing the earth's seven continents, was redefined.

Every human community then in their small worlds attempted to nurture their younger generations with reasons for these natural variations, but their explanations were often refuted by succeeding generations as obsolete or primitive ideas. Most often, the notions were replaced with empirically observed data or indisputable, verifiable evidence to show why we are what we are as opposed to speculative, unverifiable, spiritual, superstitious propositions of religious people of the past.

Empirical studies by *social* and *natural* scientists in various fields of studies have demonstrated in most cultures for several centuries that the answer to the *SED* quest must either be based on environment or genetics. Studies which are based on experiments help the observer to critically examine very close units of matter in their locations, their relationships, and how each matter's existence depends on each other.

Without clearly defined evidence, all information about nature, nurture, and divinity variation is pure speculation, fantasy or wishful thinking.

Experimentally, as an example, a researcher can conclude that A exists because of B but that C exists because of A and B's existence.

Logically therefore, A, B, and C are inherently interdependent. One cannot survive long without the others. Such empirical observations can be asserted logically as truth because there is tangible, verifiable evidence. And more so in this regard, many of the conclusions drawn here can be taken as nerve-pricking speculations and wishful thinking of a layman.

In observing human communities worldwide, one thing that shows up without dispute is the lack of progress in all communities managed *purely* by black people in comparison to nonblack communities. Is this the black man's nature, nurture, or godly design?

This creature among the *Homo sapiens* group of humans is demonstrably incapable of rapid progress because there is something suspicious in his nature which prevents him or her to receive and to process information for self-improvement and therefore unable to deliver progressive action. There is a general consensus that black communities lack progress. Conventional wisdom traces the source of the problem as environmentally related. But as previously stated, habitats play insignificant roles in organic growth and that the core reason should be sourced from the genetic content of the creature.

If the base engine or the organic components is made of inferior materials or precisely if black people are formed with low-quality cells or neurons, as it is proven to be embedded in

E. ASAMOAH-YAW

melanin pigment in a black person's body (particularly his brain organ), his actions in his surroundings would perpetually be different from nonblacks' actions by nature.

There is a long-standing belief that all men are created equal. Yes, this is true to some extent. We are all *Homo sapiens* and are therefore identical, but there are *also* natural differences and similarities in all mankind.

A United Nations–sponsored genome project declared that the genes of humankind are 85 percent identical.

The remaining 15 percent of our genes show variations, depending upon the cline type or racial group we belong to. In other words, some human groups have some genes which others don't have, or some groups have *more* than what others possess.

This 15 percent variation may sound negligible or insignificant statistically, but when the difference represents a specific unique nature or distinct characteristic of a human group (physical appearance and intellectual performance), the 15 percent variation is a huge difference and very significant.

Now, let us switch our attention to the visible variations of global creatures, animals and human beings, in their three natural main habitats—namely, space environment, aquatic environment, and land surface environment.

How do all organisms thrive in their natural habitats—their daily struggle to obtain food, water, air, shelter, and life sustaining-goods responsible for comfort, pleasure, and luxuries?

These are natural, organic physical needs without which life on planet earth is not sustainable. It can be assumed therefore that the purpose of all lives is to grow healthier, to produce our kind, to maximize pleasure, to constantly progress, and to improve ourselves and our surroundings by utilizing available natural resources to attain these goals with minimum efforts. Indeed, humans should foresee a natural need to change old ways by improving on them and creating new better ways.

To this extent, it can be asserted that those species which are alive today have been successful because they appreciate the reason why they are here on planet earth. Yes, humanity, you and I, plus all other visible creatures, have survived because we are capable of adapting to changing circumstances in our world. Simply put, we are alive today because we have been successful in our fight to remain here alive. A necessity to improve living conditions should be a human instinct.

Humans need progress because we keep increasing our numbers. Progress means advancement, growth, improvement, modernization, positive change, innovation, inventions, and evolution.

Understanding the past and present methods of production and distribution of wealth and practically increasing output to meet expanded demand should be seen as progress. Progress means changing our ways of yesterday to improved living standards. Progress is moving toward maximization of wealth easily, simply, happily, and for the better for present and future generations. In this sense, it will be safe to say that humankind has accomplished significant growth compared to other organic species, such as monkeys.

The human population has been increasing at a faster rate compared to most other species and of our recent past. Yes, we appear to have the ability to sustain population growth. But if we break down humanity into specific ethnicities, it becomes too obvious *historically* that some human groups are more progressive than others. This is an observed fact. Some possess snail-pace speed in natural progress, while others possess horse-pace speed in progress by their nature.

Some have the natural desire to remain stagnant, conservative, and actually survive on the innovations and creative ideas of others. This observation is fundamentally natural and actually has very little to do with environment. It is the creative minds of humans which initiates environmental progress. In other words, it is *total amount* and *quality* of information captured by individuals' brain organ, when processed efficiently, that cause all human actions. The habitat may incite the indigenous people, but it will primarily require a human's natural intellectual capabilities to actually initiate effective practical changes or progress. What counts most therefore is the core quality of the human brain organ or the body's basic constitution which senses and controls environmental information.

The brain captures information and processes them for the body to act or react. That counts first. Everything else is an added factor.

It will be later discussed in detail in this discourse to show why some organic matters, including some human beings, are naturally unprogressive and naturally exist as parasites to other progressive humans.

All types of people naturally think firstly of survival needs, which include eating, drinking, breathing, and hygiene. Some people, however, do not just do all the four above, they eat more *good* food—more vegetables, more fruits, more fish, etc. They drink good water, not just water. They breath good air, not just air, and consciously live in hygienic environments, not just living anyhow. Yes, all people make effort to eat, to drink, and to breathe because all of us need to stay alive. But some merely eat to satisfy hunger. They drink to quench thirst, regardless of nutritional values. They breathe air with no regard to its purity and care less about the sanitary conditions they live in.

*Most* black people per se do not think further than that which would sustain their immediate personal life. *Most* blacks are concerned with surviving today. Tomorrow will be taken care of by God or by itself. *Most* blacks do not plan ahead.

*Most* blacks don't have a one-year plan, a five-year plan, a ten-year plan, or a long-term plan. *Most* blacks therefore float like a leaf on a river surface and are ever ready to be blown by wind and river current by their god's will. Most blacks are not critical thinkers—too naive, too simple, too noisy, too lousy, too easily convinced, and too lazy.

*Most* black people are afraid to take risks in anything and therefore do not achieve success in anything. Most blacks don't know that success and risks are natural partners in life.

*Most* blacks are afraid to take risks because they are scared of losing. They don't know that losing is the first lesson toward progress. "Nothing ventures, nothing gained" is an ancient dictum.

*Most* blacks identify themselves with foreign ancient religious faiths, such as Christianity and Islam, and pray every day for their god's intervention in their lives, although historically all black cultures possessed indigenous religions.

Most blacks believe there is something called miracle, and they pray tirelessly and hope for miracles daily. Most blacks are convinced that the foreigners' god in the Holy Bible and Quran will provide for their needs at God's own chosen time, and they are prepared to wait indefinitely.

E. ASAMOAH-YAW

*Most* blacks do not want to think beyond faith. While most nonblacks dream of inventing the unthinkable, most blacks dream of being chased by the devil.

*Most* blacks are poor and know not how to improve themselves and therefore live in poverty. *Most* blacks think and behave selfishly and are too comfortable with mediocrity. *Most* blacks do not understand sanitation. *Most* blacks are decadent and live in filth.

Mosquitoes have been killing millions of people in black African communities for centuries. A means to control it (eradicate or neutralize the parasite) is not undertaken by blacks but rather by nonblacks' governments, universities, laboratories, and nongovernmental organizations. Malaria medicines are created and manufactured by nonblacks.

The concept of unity, society, community, or being a group of common people with a common destiny is rarely part of black people's everyday concern. Why? Doesn't it mean there is something seriously naturally wrong with fellow blacks? We live it. We see it. We simply don't a give a damn! What happened to our humanity? If we are really qualified to call ourselves humans, like most nonblacks, what is the black person waiting for?

*Most* blacks are ever ready to blame governments, Satan, witches, invisible entities, and others such as their former colonial masters, slave masters, and white people in general for their natural weaknesses and faults rather than themselves. If most black people are smart enough than most nonblack people and also know whom to blame, why can't they outsmart nonblacks in every field of human progress?

Why can't black people do something to prove their worth if they are really worth more than my estimation? Most black people are too far behind human development, except pure in buffoonery.

Yes, I am a pure Negro and supposedly fully educated (BSc in economics from the University of London, 1976 alumni). Being a black man for seventy-four years (as of the year 2016)—half of these years were spent in Europe and America— clearly teaches me that my race of people have serious *natural soluble problems*.

*Most* blacks do not question status quo and remain very conservative and ambivalent. *Most* blacks are casual, transient, and out of focus. Yes, I am a pure black person and Ashanti by heritage. My ancestors are genealogically pure melanin-rich Africans. I know what I am talking about. I have lived over half a century with many of my kind and with millions others, including East Indian blacks, Australian Aboriginal blacks, diaspora black Americans, West Indians, and black European settlers.

I have also socialized and lived with nonblacks of all types for thirty-five years, and be assured I am conscious of what I am comparing.

There is something definitely wrong with my fellow black people by nature. Sadly, most blacks naturally know not what they know not. Above all, most black people do not like reading books; hence, their knowledge is limited mostly to what they see and hear.

Black people hate books because they don't want to know anything new that can challenge information already stored in the perverted brains.

Most highly educated black people fail to apply their acquired academic knowledge efficiently in the practical world. The knowledge has been acquired primarily because of the academic title it carries (i.e. bachelor's degree or, master's degree or doctorate degree). Most black people are duds when it comes to defending their academic titles. Most fail to originate new ideas and hence perpetually depend on obsolete concepts in their fields of study.

He or she will be proud of the academic laurel forever, and only a few contemplate even to write a pamphlet about the faculty. His or her neurons are essentially capable of mimicry and lack useful imagination.

*Most* of the *few* articulate blacks think their "weaker" difference is caused by their environmental circumstances or factors, such as cultural norms.

While most black people would agree to most or all the above allegations, they do not accept nature as the culprit of their demise. Most blame or retain the obsolete notion of the environment as the main cause of what they are today.

Again, if most environmental issues are humanly resolvable, can't black people wipe out their defects by environmental means all these centuries? Obviously, blacks are waiting for nonblacks to come out with a permanent solution to correct or to purify their naturally faulty neurons.

Although I don't have scientific proof or formal statistical evidence to support these claims, I have lived it, have read about it, and have practical experience of it. I am convinced all the above have a lot to do with a black person's natural makeup. Black people's nature or biochemistry is responsible for black people's lack of full humanity.

Our inferiority must be genetic, and I hereby throw a serious challenge to geneticists, chemistry and biology specialists, neurologists, psychiatrists, and the entire scientific community to revisit their laboratories and search for the culprit cell or the organic matter responsible for black people's natural neurological weaknesses. The answer could be hiding in one of the cells

E. ASAMOAH-YAW

which have the capacity to produce melanin. I say melanin because it is the major natural substance which black people have more than any other human species.

There ought to be a natural correlation between melanin and the brain capacity of a black person for receiving, processing, transmitting, and retaining information. Most black people exhibit wide negative differences in these areas compared to nonblacks.

I challenge every reader of this work to look around your immediate self and count all material objects. How many would you say were created or improved by a black person? You can hardly get one or two. And it would not be a coincidence.

It cannot be the environment or some kind of natural coincidence that *most* people with very low melanin, such as Caucasians, perform much better than *most* people with abundant melanin, such as Negroes.

If intelligence can be defined as the potential ability and capacity to acquire, retain, or process and then apply experience, knowledge, understanding, reasoning, and judgment in coping with new experiences and in solving problems, then I can conclude without hesitation that *by nature, most* black people are less intelligent. Only a few black people are intelligent *by nature*. Also by nature, *most nonblacks* are intelligent. *A few nonblacks* are less intelligent by nature.

If we compare human aptitude in life, it shall be seen that in general most black people lack the mental capacity and ability *to acquire* knowledge, *to retain* knowledge, *to process* knowledge, and *to transmit* knowledge or information outside the body.

I am aware of the provocative nature of these statements. But as a patriot of my race, I am confident that black people can be fine-tuned to perform competitively in modern science.

Life is not just about surviving. It has more purposes than that because as long as our life sustenance is fulfilled, it is most likely we shall live to see tomorrow, next week, next month, and years ahead. And beyond our usual mere struggle to survive, humanity needs proper shelter, proper clothing for protection of the self, and security of the acquired wealth, whatever its value may be.

Humans need to recognize and appreciate accomplishments when there is exhibition of excellent performance by ourselves. Most blacks mostly thank their god for their intensive hard work and achievements. They know not how to appreciate their personal efforts—that every hard work carries its unique proportionate reward. The harder you work, the greater your rewards. The lesser hard you work you do, the poorer your reward. Hard work and success must be established in the mentoring of the youth or the have-not to copy and to pursue goals. Thanking one's god diminishes human potential and inclusiveness.

While *most colored people* will be satisfied with the most basic shabby clothing and basic housing accommodation with no kitchen, no bathroom, and no toilet, *most nonblacks* will show concern when these facilities are absent. Most nonblacks think beyond mere basics in life.

It is normal and very common in Ghana—in the year 2016, for example—for a whole township of 200 houses with 2,000 people to share one toilet facility for males and one for females. Besides this public facility, some citizens comfortably go to toilet haphazardly in gutters and under trees. Open defecation is common in all cities and towns in most black African countries. It appears that open defecation is indeed "not a big deal" to most colored people's communities in Africa, some parts of Asia, and some South American colored societies as well. These are some examples of uncivilized behavior of most black people which makes white people in general classify blacks as lesser humans.

Most black people for centuries have lived recklessly and hopelessly and see no reason to change their mind-set because such is their nature.

A well-trained dog or cat, for instance, will excrete in a designated place only, and so will most nonblack people or civilized human beings. A trained dog will wait cautiously at a pedestrian crossing and look left and right to ensure safety before crossing a road. Most blacks in black communities do cross roads anywhere and anyhow, disrespecting road regulation. Lawlessness, recklessness, and lack of common human ethics are generally the norm among most Negroes by nature.

Sadly, these black citizens see nothing wrong about their unhealthy behavior. Most of these citizens are comfortable, for instance, with used clothing and appliances imported from nonblack countries. This is so because they are cheaper, easily available, and supposedly have better quality than locally manufactured goods if the alternative is locally available.

Secondhand manufactured goods, such as motor vehicles of all kinds and used clothes (such as underwear, bras, socks, shoes, pants, shirts, blouses, headgears) are traded, preferred, and cherished by most black people in Third World countries. The popular excuse is that most blacks cannot afford to pay for brand-new ones or that blacks do not have machines and know-how to produce these goods.

Another big question is, why not the reverse, if black and nonblack people are naturally the same? The reason that blacks cannot afford new ones or that used ones have better quality than new goods is nonsensical. Why is it a fact that most black people are impoverished and undeveloped, or why are most nonblacks rich and developed? Why have we always blamed the environment rather than our nature for common weaknesses when we know for sure that we are potentially naturally richer than most nonblack countries?

Black countries, for example, possess more natural resources and hence are environmentally far richer than most nonblack countries. Isn't it surprising that blacks naively blame the same environment which is ever-ready to be exploited as the source of their low esteem? What prevents black people from exploring their land resources to invent or manufacture goods for their own use and, if possible, export the surplus for extra income?

For lack of intelligence or foresight, black people would rather wait or beg for capital and technical know-how from nonblacks to come to explore natural resources in their own backyards. Black people's nature is demonstrably below most nonblacks' natural standards. Generally, blacks are unfit. There is a missing human trait.

No black African capital city has a centralized sewage system as of the year 2016, and this is not a big deal to black African citizens and their leaders.

Fifty people in a household will be comfortable to share one kitchen in turns or, in many cases, will have no single designated place for cooking and will have only one enclosed place for a shower for both sexes and run it in turns.

Each family group would for generations cook comfortably in front of the compound house with charcoal or firewood and not conceptualize a need to progress an inch beyond what they have inherited from their ancestors.

People, adults and infants, literally urinate everywhere and throw away trash haphazardly inside and outside the houses. They wait for rainfall to wash the filth away into rivers they depend on for drinking and for all household purposes.

To most black communities, this is not a big deal to worry about because when they fall sick or die as a result of filth, a witch (oldest female) in the family will sadly be blamed for it.

*Most* black communities do not have pipe-borne water supply. And those who are privileged to possess pipe-borne water supply (created by former colonial white masters) do not see a need to maintain the facilities. They carry pots on their heads to fetch water from nearby rivers for drinking and food preparation. They use soap to bathe in the same rivers. They wash clothes and feed animals in the same rivers. Rivers in these communities are polluted with filth, human waste, hazardous industrial materials. They deliberately pollute their environment, suffer the consequences, and blame governments for not ensuring cleanliness. At the same time, they pray every day in churches and mosques for God's protection against illness. When people fall sick in most black African communities, they seek divine healing first from a local church pastor or a fetish priest. They seek professional medical attention or go to hospital when conditions become very critical. This is very common in Ghana and all West African countries even in the year 2016.

Water as a life-sustaining element is meaningless to most Negroes.

Black people's overall behavior when compared to nonblacks, especially compared to most Caucasians, is totally different and shamefully very inhuman. Yes, it is our nature!

Colonial slave masters (white Europeans) established pipe-borne water supply in most district capitals during the later part of the 1920s through to the 1960s when they finally conquered and took over control of the Negroes completely. Public standpipes were created for general-public use in big towns of several communities who could not extend pipe water into their houses. But a decade after granting political independence to indigenous black populations, all institutions, including water supply systems, electricity-generating plants, postal and telephone communication systems, roads and transportation systems, education systems, central and local government systems, judicial systems, plus several institutional systems broke down and were unmaintained or destroyed.

All these facilities began to deteriorate because maintenance as a code of conduct is naturally not part of most black people's faculty. The Public Works Department, which was established for maintenance of all public facilities, was neglected and eventually closed without a replacement. Public structures in most black African countries have deteriorated or have been abandoned, demolished, stolen, or sold to private developers. Black public officials and politicians of the day share the proceeds with impunity.

As mentioned several times earlier, the black person in general and everywhere has a natural inclination to crave contentment for mediocrity, simplicity, laziness, selfishness, immaturity, and sheer stupidity.

*Most* black people are naturally complacent and always try to find someone to blame for their own irresponsible acts. Why?

It is not in most black person's nature to conceive an idea to alter the old ways of living or of doing things better than those inherited from ancient ancestors. His blackness or his more-than-necessary melanin granules in the brain tissue, the skin, the hair, and numerous other locations of the body.

I mean, his main genetic difference should be held accountable. It *cannot be* his or her habitat or environment which molds him or her.

Some of his trillion body cells need to be investigated by research scientists. Because in this day and age (the year 2016), it is highly convincing to say that natural traits of many organic matter can be altered or reconstituted.

Gone are the days when everything natural was seen as untouchable, irreversible, irreparable, or God made.

It is evident that *most nonblack people* do not act senselessly as *most black* people do. *A few nonblacks*, admittedly, naturally possess these black abnormalities. And even these few naturally superior blacks who can pioneer progressive changes within the race have not yet shown awareness of these black persons' natural deficiencies. Why? They are in self-denial and appear to suffer from alienation.

A black person has a large amount of melanin deposit; similarly, plants' and animals' chemistries, according to scientists, also contain large quantities of this melanin substance.

By melanin implication alone then, black people as a type of organic matter or human species are closer to plants and animals than we are to *most* nonblack people. And again if *most nonblack* anatomies contain insignificant or a very small deposit of melanin in their hair, eyes, and nipples, would it not be *sufficient evidence* of most nonblack person's nature as being naturally different and consequently *more progressive, innovative,* and *inventive* than most blacks?

*Progressive actions* in science and technology in most nonblack communities are classic evidence of these conclusions. Most nonblacks understand nature better than most blacks because most blacks see no reason to interfere with nature and therefore prefer to let nature take its own course.

# CHAPTER 4

# The Melanin-Loaded Human

## Melanin

WHAT IS IT?

1. Dictionary definition:

(a) "Greek root word (*melan* or *melano*) means *black* or *dark*. (i). **Mel-a-nin**; a dark brown or black animal or plant *pigment*. (ii). **Mel-a-*noid***: A pigment (as one contributing *esp.* to the yellow color of the skin) that is a disintegration product of a melanin" (source: *Webster's Ninth New Collegiate Dictionary*).

(b) "It is a dark biological pigment (bio-chrome) found in skin, hair, feathers, scales, eyes, and some internal membranes. It is also found in the peritoneum of many animals; an end product during metabolism of the *amino acid* and *tyrosine*" (source: www.britanica.com/scien. Encyclopedia).

(c) "Any of the various black, dark brown, reddish brown, or yellow pigments of animals or plant structures (skin, hair, choroid, or a raw potato when exposed to air)" (source: *Merriam Webster Dictionary*).

2. Scientific definition:

(a) "A black or dark brown pigment that occurs naturally in the hair, skin, and iris and choroid of the eyes" (source: *Mosby's Medical, Nursing, & Allied Health Dictionary*, fifth edition, ISBN: 0-8151-4800-3).

(b) "It is a natural substance that gives color (pigment) to hair, skin, and the iris of the eye. It is produced by cells in the skin called *Melanocytes*" (source: <u>www.ncbi.nim.nih.gov/pubr</u>).

## Melanocyte

It is a body cell capable of producing melanin. Melanocytes are distributed throughout the basal-cell layer of the epidermis and form melanin pigment from tyrosine, an amino acid. Melanin granules are then transferred to adjacent basal cells and to the hair.

Melanocyte stimulates hormone from the pituitary and controls the amount of melanin distributed throughout the body.

*Three major* types of melanin are produced by melanocytes (www.news-medical.net/health/whatismelanin):

1. *Eumelanin.* It is found in the hair, skin, and dark areas around the nipples. It is particularly abundant among black people populations, and it provides black and brown pigments to the hair, skin, and eyes. *When eumelanin is present in small quantities, hair may be blonde.*

2. *Pheomelanin.* This melanin is also found in the hair and skin. It provides pink and red colors, and it is the main pigment found among red-haired individuals. It is not as protective against UV (ultraviolet) radiation as eumelanin does.

3. *Neuromelanin.* This melanin is found in different areas of the brain, and loss of this melanin may cause several neurological disorders.

## Miscellaneous Scientific Terms

*Gene* is a biologic unit of genetic material and inheritance. It is currently considered to be a particular nucleic acid sequence within a DNA (deoxyribonucleic acid) molecule that occupies a precise locus on a chromosome, and it is capable of self-replication by coding a specific polypeptide chain.

Genes are broken down into *twelve major groups*:

1. complimentary genes
2. dominant genes
3. pleiotropic gene
4. recessive gene

5. regulator gene

6. structural gene

7. lethal gene

8. sublethal gene

9. supplementary gene

10. wild-type gene

11. mutant gene

12. operator gene.

Note: It is important to know that the presence of lethal gene and sublethal gene *causes or impairs the functioning* of the organism that possess them. However, they *do not cause* its death.

*Genetic engineering* is the process of producing recombinant DNA so that the genotype and the phenotype of an organism can be *altered and controlled*. Enzymes are used to break the DNA molecule into fragments so that genes from another organism can be inserted and the nucleotide *rearranged in any desired sequence.* The technique can be applicable to higher organisms with the possibility of controlling and eliminating genetic disorders and malformations in humans (source: ibid., *Mosby's*).

*Genotype* is a genetic material of an organism.

*Phenotype* is the visible traits of organic matter.

*Homo* means "of the same."

*Genus* means "a subdivision of a family of animals or plants."

*Homogenous* means "having a likeness in form or structure as a result of a common ancestral origin."

*Homo sapiens* means "scientific term for the genus and species identifying humans."

*Homogeneous* means "(Gk) consisting of similar elements or parts: Having a uniform quality throughout."

*Race* means "breeding stock of animals; a family, tribe, people, or nation belong to the same stock, class or kind of people unified by community of interest, habits, or characteristics: a division of mankind possessing traits that are transmissible by descent and sufficient to characterize it as a distinct human type: inherited temperament or disposition" (source: *Webster's Ninth New Collegiate Dictionary*).

I am informed, however, that some people do not possess melanocytes in their blood chemistry or that they have the cells but do not produce any of the melanin pigments. It is one of those natural organic disorders.

These people are called *albino*. Black and nonblack albinos do exist in every human kind.

Albinism is a congenital hereditary condition characterized by total or partial lack of melanin pigment in the body. By implications, complete elimination of the culprit cell (melanocyte) will bring some ailment to the body. Apart from melanin, the cell is known to produce several functions in blacks, nonblacks, animals, and plants. My concern is its management in black humans. It can be controlled through genetic engineering.

Black people possess too much of the cell's produce. Its production of excessive melanin is the major known natural difference between black and nonblacks. Why does a normal black person's body, especially in the brain tissues and neurons, contain more melanin pigment than nonblacks? I hope neurologists and genetic engineers will examine critically the *neurons on both sides*, recognize the black difference, and do something about it.

In the book *The Bell Curve*, published in 1996, an attempt was made by the authors (Herrnstein and Murray) to propose the idea that a natural intelligence difference exists among racial groups in America and that the first step of bringing equilibrium is by acknowledging its reality, but their conclusion was not universally accepted. Several experts on both racial divide condemned them as being racist. The phenomenon of natural racial inequality is not new. The question pops up frequently, especially among ordinary black people. The not-so-ordinary blacks, such as the highly educated ones, do not hesitate in concluding that if most black people strive as hard as necessary to focus on desired goals, everything is possible regardless of racial heritage.

But one thing these advocates choose to ignore is the historical fact that most black people have no defined life ambition and are out of focus. And they fail to strive as hard as necessary to achieve defined goals.

They are incapable of maintaining positive focus on defined goals because there is no natural burning desire to struggle to succeed by any means possible.

They mostly give up at the point of critical thinking.

This black people's behavior is conducted by the organ of thought, the brain, which is responsible for all ideas or information so far captured through exposure by its human owner.

# CHAPTER 5

# Brief History of Mankind

HISTORY IS A story of the past. It is always presented in two forms. It is either a written history or an oral history. Whichever way it is tackled, the story must always have a beginning and an end. The storyteller must define the perimeters of the object in question and tell it as it is.

There are many inherent problems in telling or writing histories, especially when the author is not an eyewitness or a participant of the story during the time the event occurred. Every story contains margin of errors because the information is given from the teller's perspective;

the choice of words, sentences, meanings may be different when told by another.

Such errors as in chronology, place, facts, and circumstances are some of inevitable distortion elements in all histories. Narrators as well as readers or listeners must be rational, cautious, and curious throughout.

The planet earth, the world, and its contents have histories.

Stories of events vary in many ways depending on who the storyteller is and for what purpose the story is being told, but generally, academic historians' versions are considered authentic. Again obviously, the absence of eyewitness accounts of how and when the earth came about makes mankind rely on scientific explanations, such as that of anthropologists' estimates of its nature, which are imperative and indisputable but often a bit skeptical.

The planet earth is estimated to be about 15 billion years old. The earliest known fossils or human remains that can be matched or identified with modern humans are those of *Homo habilis*, which dates from between 1.8 million and 1.2 million years ago. The next known evolutionary stage of humans was *Homo erectus*, who first appeared about 1.5 million years ago. The earliest fossils of our own species, *Homo sapiens*, dates back from *c.*400,000 to *c.*250,000 years ago.

An apparent side branch of it, the *Neanderthals*, is known to have existed in Europe and west Asia some 130,000–40,000 years ago.

*Fully modern humans*, the *Homo sapiens sapiens*, first appeared *c.*100,000.

Several years ago, opinions were divided as to how exactly the species emerged, but by 30,000 BC, this *Homo sapiens sapiens* (the double *sapiens*) was the only surviving hominid. Meanwhile, some parts of Africa retained the *homo sapiens,* who had been evolving slowly for thousands of years. It is not yet established if this species will eventually be at par with the *homo sapiens sapiens*.

That black people are considered the first inhabitants of planet earth is currently the conventional wisdom.

We are further informed that these creatures lived somewhere in the southeastern part of Africa. Their offspring migrated and populated the entire planet through Asia, Australia, Europe, and the Americas.

By evolutionary process, black people's ancestors, *Homo erectus*, dominated the entire planet with several accomplishments, including inventions of the bow and arrow, agricultural tools, pottery, taming wild animals (such as dogs, horses, goats, sheep, chickens), and later, the wooden wheel and several others. They also discovered techniques in smelting gold, iron, silver, and other mineral resources.

Through oral history and modern scientific instruments, several documentations have been created and authenticated to support the idea that the early settlers of planet earth are Negroes.

Human evolution from monkeys to *africanus*, to *Homo habilis*, to *Homo erectus*, to *Homo sapiens*/Neanderthal, to *Homo sapiens sapiens*.

The third image, *Homo habilis*, shows use of primitive stone tools. The fourth image, *Homo sapiens*, evolved with improved tools to the present-day sophisticated *Homo sapiens sapiens*.

# African Origin and Track of Humans by Anthropologists

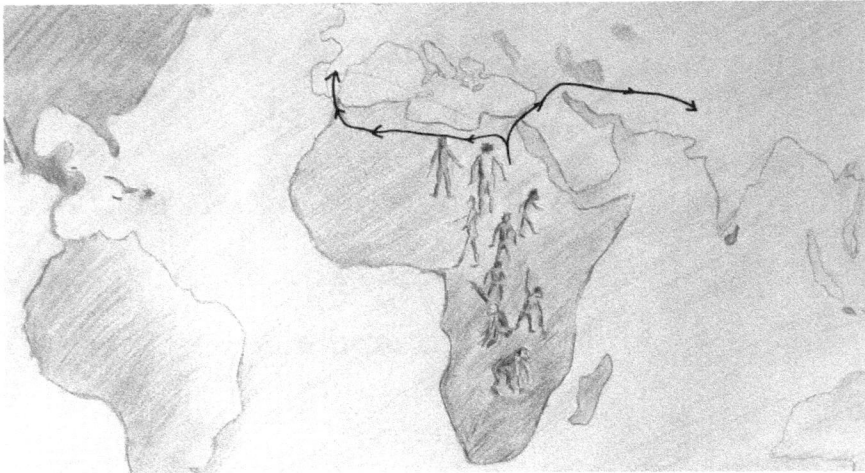

South-eastern Africa as the source of First humans on planet earth. This is how we evolved as animals into who or what we are today as homo sapient sapiens humans; and how we spread across the planet.

According to the above illustrations, darkskinned people were actually the first inventive, progressive, and well-organized kind of humans.

The black-skinned person was the mother of all and was really the most intelligent when compared to other living organisms. There were no other humans to compare with. Today there are varieties of descendants from this very first human; hence, comparisons can be made among all human species whose ancestral heritage is traceable only from black people.

As already compared in several pages earlier, the nonblack humans are completely different from black people in almost every conceivable way apart from the basic natural human anatomic characteristics. Despite the marked differences, there are also some natural similarities between racial groups.

Some experts trace the emergence of the blacks from about 600,000 years ago when humans did not exist. However, it is not yet established if this creature simply evolved from animal species or how this primordial creature was formed, with what, and by whom.

The popular answers to these questions differ from culture to culture. With over 7,000 current world cultures, there ought to be over seven thousand answers.

However, scientifically, all humans are believed to have evolved from the combination of water and other organic matter. This also evolved into algae or bacteria over the years and, after a series of multiple transformations, gradually grew into living beings in the aquatic world, such as fishes of various types. Over the years, water creatures multiplied in numbers and sizes. Some migrated to land surface and finally made open lands their homes or habitats.

All these happened because of the availability of intangible energies, like light from the sun and the wind or oxygen, carbon, and nitrogen. *Homo sapiens* to *homo sapiens sapiens* or chimpanzee-looking humans into humans on two legs, two hands, and hairless skin.

Now let's switch from prehistoric and cross over to ancient history to the Middle Ages records of the black person. Commencing and comparing from the fourteenth century, we see that prior to medieval days, the world has changed a lot. All the continents are occupied by unique types of racial groups: Asians, Orientals, Caucasians, Hispanics, Africans, and others. Each continent is being dominated by a specific racial group. Then after a search through each landmark's history, especially the black person, each group plays a major role within the confines of its own continent and in its relationship with ethnic groups in other continents.

A common behavior which prevails in all the seven continents is a constant internal struggle for dominance of one culture over the neighboring cultural or tribal group. The primary purpose of earlier wars was acquisition of land or territorial expansion for sheer feeling of pride and dominance. The earlier territorial expansion was not for economic reasons. Who was militarily powerful was the prime motive.

In *Asia*, for instance, the Chinese dynasties were fighting among themselves. They were the first to invent gunpowder.

In India, each of the over two hundred tribal groups were fighting each other to prove their superiority over the others.

The Mongolians were killing each other to show ethnic superiority.

The Japanese had their ambition of global dominance beyond their limited territorial island boundaries.

The Australian peninsula, with the numerous islands surrounding the continent, lacked well-defined governmental systems to manage their own progress. Inter- and intratribal wars guaranteed stagnation for centuries because of pride.

In *Europe*, the situation was even worse. The Roman Empire was the longest and largest single ruling authority the world has ever experienced.

Its crackdown ignited monarchical wars in Europe: struggles for unity in the Italian city-states of Florence, Milan, Genoa, Bologna, Pisa, and Sicily; the Roman holy wars and political power struggles; the Portuguese and the Spanish adventurous wars; the conquest of Normandy, the Austria-Hungarian imperial struggles; Oliver Cromwell's challenge of the English aristocracy, resulting in parliamentary democracy; the English, Welsh, Scottish, and the Irish Union infractions; the Russian czarist integration or imperial conflicts leading to the Bolsheviks Revolution, Marxism, communism, etc.; Bismarck, the Weimar Republic, and the imperial wars of Germany, Greece, and

Turkey; the Balkans, Serbians, Turks, and later, the Ottoman Empire's ambitions to conquer the entire Europe; disintegration of the entire continent of Europe into several alliances (the Triple Alliance, the triple Entente, French and English and Russia Alliance); European fascism; World War I, Nazism, and World War II; atrocities and colonization of the Americas by Europeans; division of the Americas into colonies belonging to European rival countries.

In continental *America* proper, scores of indigenous cultural communities had been at war with one another like the rest of the world. The genocide of the continent populace by Europeans resulted in complete extermination of the indigenes, which includes the Inuits, the Incas, Mexicans, Takomas, Toledos, Aztecs, Azilin, and many others.

In ancient Africa, the dominant populations of the east and southeastern province, where ancestors of the original *Homo sapiens* is rooted, were engaged in wars to alginate those who did not speak the same language. They emerged and increased and later spread across the entire continent, always struggling to dominate weaker groups.

They organized into many linguistic clans or groups and formed principalities, kingdoms, and dynasties.

Many weaker ones were conquered by the few imaginative and resourceful leaders who managed to organize their people to build and acquire defensive and offensive weaponry to maintain territorial superiority.

The kingdoms of ancient Ghana, the Songhai Empire, the states and kingdoms of the forest and the Guinea coast (such as the early Akan states, Asante Empire, Kingdom of Dahomey, the Oyo civilization, the Yoruba, the Hausa, Katzina, Zaria, Chadic, Bornu, Soninke,

Mandinke, Serer, Susu), and many such kingdoms in southern Africa once *in ancient times* demonstrated that *Homo sapiens* black people could manage themselves.

The portrait of Africa—in particular, in the Middle Ages and medieval period (AD 500–1400) changed completely from ancient times when the *Homo sapiens sapiens*, the nonblacks, came in contact with black Africans (*Homo sapiens*). Bear in mind that this visitor evolved from the same single *sapiens* into the double *sapiens*.

Over several thousands of years of departure from east and southeast Africa to the Middle East, Europe, Asia, etc., they returned completely changed from top to toe.

Hair texture and color, skin color, and overall complexion, plus total physiological image change, made the former cousins a complete alien. Indeed, both humans were beyond recognition to each other apart from the normal human appearances of similarities, equalities, and differences.

Between AD 1300 and 1950, the whole world had passed through unprecedented evolutions and revolutions in every aspect of human life. This came about because *dominant cultures* among humankind, with their superior body chemistry, manufactured powerful weaponry and had creative ideas to conquer or colonize other humans with weaker cultures and low-quality biochemistry, such as blacks.

In other words, nonblack people, especially Caucasians, successfully conquered, suppressed, dehumanized, and dominated colored people for over 650 years because people of color did demonstrate over the years that they could be suppressed and be treated as subhumans.

It appears from this point of view that black persons' nature as humans is either not completely developed or fundamentally developed with less efficient natural components.

Modern black people resemble the earlier human species, the *Homo sapiens*, who lived in east Africa between *c*.400,000 and *c*.250,000 years ago. Black people then did not behave like the nonblacks or the *Homo sapiens sapiens* of *c*.100,000 years ago.

It seems the black people are still developing, possibly, from the original *Homo sapiens* category of human species toward the *Homo sapiens sapiens*, the refined human group.

All humans do develop. Some develop much faster than others. Development among blacks has been unusually very slow because growth and development is accelerated when there is a compelling natural necessity for it. Black people's domain has always contained everything they need for survival. Natural resources abound everywhere, and they need not struggle too much or think critically to make ends meet. There is always plenty of sunshine and rainfall; every bit of the land is potentially rich to grow food to feed the family and beyond.

Ostentatious living and affluent lifestyle was and still is alien and unthought of to a typical Negro. The black man did not need electricity and its accessories, like refrigerators, air conditioners, light bulbs, radios, televisions, etc. Basic, simple living is in a black person's nature. He saw nothing wrong at all with that.

And if *necessity* is really the mother of inventions, as the popular saying goes, what would be the motive for a black person who lived in that part of Africa at the time to imagine creating or manufacturing something new? A straw-and-mud hut for shelter, shabby clothing to cover some private parts of the body, clay utensils, and ordinary human needs just to stay alive and healthy is awesome enough to fulfill a black person's needs.

Again on necessity or needs as a conductor of rational action, the black person cannot be judged by standards of other humans, he sees no reason to change. His offspring migrated from their primordial habitat in the east and southeastern part of Africa to the northern regions of the world and, centuries later, through the Middle East to Europe, Asia, Australia, and the entire world all because of population growth, overcrowding, and unanticipated social problems between the Neanderthals and the *Homo sapiens sapiens*; humans had plenty of reasons to not only change but much faster. Survival in the tropics and in the temperate zones is not the same. Necessities for suitable shelters, clothing, food, and indeed, every survival need were scarce or not available at all in that part of the world.

The unpredictable climatic conditions compelled this newly evolved species to explore beyond their ordinary needs in order to survive. Their increasing numbers through procreation increased demand and supply for food and consumer goods; it created compelling reasons, especially Caucasian humans, to be as innovative as possible.

Circumstances in their adapted environments compelled them to cooperate harmoniously within each cultural group for the survival of all. This led to intense struggle to expand territorial boundaries, followed by wars among nations in Europe for centuries until the last decade of the twentieth century.

Thus, planning ahead of time, being creative and adventurous has always been imperative with this type of humans outside the African continent.

Unlike the original *Homo sapiens*, who decided to remain in Africa just because life was easy, simple, and less stressful, their offspring, now the *Homo sapiens sapiens* (nonblacks), had to think beyond the ordinary. There was a need for utilization of their natural resources to manufacture accessories and equipment to substitute manpower and for creative ideas to protect and improve physical conditions and weapons against possible aggression.

They wandered around their habitat and neighboring lands, rivers, and seas, exploring every unknown matter for everyday and future use. These are the types of humans I describe here as nonblack persons.

The original human, the *Homo sapiens* (black persons), had no reason to think beyond the ordinary. Everything needed for their survival has been naturally available at all times with minimum effort to acquire them. Why should the black person be innovative, progressive, perform, or change their mind-set when the day's comfort is known to be sufficient? Besides, is the purpose of life not simply to pursue happiness and maximization of comfort as perceived by the individual?

Judging by the above brief history of ancient and medieval people in their cultural confined environments, very little can be questioned about black people's lack of progress beyond ordinary survival of human needs. But measuring the black person as a member of the current human race, where formal visible and invisible barriers of all kinds of the past are erased or exist in shadows, thus making the globe look like a small village, it is doubtful if there can be a justification to defend the shortsightedness attitude of most black people.

Nonblacks have advancements in agriculture and food production, land resource exploitation, creation of new and improved manufactured goods, general technology in every field, communication systems across the planet (on land, rivers, and high seas), air flights and wave bands, and exploration of many unimaginable spheres of the modern world. In comparison, the insensitive nature of most blacks in these areas in their immediate and remote surroundings plus his general apathy compared to nonblacks make black people's nature exceptionally puzzling.

A relevant question that may be asked is, if both blacks and nonblacks were originally from the same parent and therefore possess the same body chemistry, a significant change in either of them would necessarily have to be caused by their different environments in collaboration with their nature. And as already stated earlier, over a long period of time, every living organic matter needs to adapt to environmental forces to remain alive. In this regard, how do we account

E. ASAMOAH-YAW

for black people's snail-paced insignificant change and perpetual decadence and nonblack people's continuous progress?

I can challenge anyone for the sake of argument that if you bring all black people to settle in Europe and, at the same time, transport all Europeans to go to settle in Africa for only two years, the European settlers in Africa will change the African continent completely into a progressive paradise, while the African settlers in Europe will mess up European countries such that the African settlers will run on their knees to beg the Europeans to come to their aid. In just a two-year experiment, all public utilities will come to a halt. Factories will stagnate and eventually close down. Infrastructure will collapse for lack of maintenance. Scientific researches and development from day one will be frozen indefinitely. There will be complete anarchy.

The best they know will be demonstrated: enjoyment in frivolity and entertainment, corruption at every level, fulfillment of private interests at the expense of public interest. There will be lawlessness, political partisanship, lack of planning and direction and control, and general indiscipline. The African Europeans will eventually turn around to blame European Africans for not teaching them how to live as real *Homo sapiens sapiens* or human beings. The European settlers in Africa will have to return to Europe to colonize the Africans.

It is the nature of the African blacks; most of us are incapable of common human rationality due to faulty neurons in the body chemistry.

# CHAPTER 6

# Dark-Skinned Person: Negro

WHO IS THIS person? What are the accepted *natural differences* between Negroes and non-Negroes? How significant are the natural differences? Is it just the differences in the skin, hair, face, and the entire physical body, or is it something more profound inside the physical body? Why is this person characterized with all the above names by the rest of the human race? Is it because of these natural cosmetics, overall behavior, or the content of the genome?

According to experts in biochemistry, all humans are equal to about 75 percent biochemically. Some experts say all humans are equal to about 85 percent biologically. The 25 or 15 percent difference in chemistry or biology among all humans appear to be an observed natural fact, and this means that some ethnic or racial groups of humans possess more or less certain natural elements which others don't or have very small amounts of; therefore, this indicates the existence of natural similarities, equalities, and differences.

A Negro is commonly known as a member of mankind distinguished from members of other races by the usually inherited genetic and physiological characteristics which have nothing to do with geographical location, language, or culture. Brown- or dark-skinned persons, or black persons, are normally grouped together under one umbrella or title known as a racial or ethnic group. It is indisputable that a dark-skinned person is a human being like the rest of humankind, only that the body's natural color, hair texture, and facial features make the black person a distinct kind of human by visual appearances. This sort of classification or variation is prevalent in all organic matters or living things.

As explained before, all species come in different types organically, anatomically, and physiologically; for example, domesticated and wild dogs (foxes and bulldogs) are either canine or vulpine mammals. They belong to one family.

They are equal and different yet similar by nature.

There are indeed several hundred of dogs with many distinct natural similarities in sizes, images, behaviors, and performance efficiencies, just as there are in humans. It is greyhound dogs' nature to excel in running. Bulldogs cannot compete with greyhounds no matter how much they are trained to compete on a race course.

Every kind of dog species has its unique natural quality. Black people's differences must be seen, firstly, as our nature and secondly as our nurture. There exists a natural racial difference both internally and externally.

Most black people's ancestry is commonly *traced* from Africa, south of the Sahara Desert, although there are dark-skinned people with similar organic characteristics in Australia, some parts of India, the Americas, and some parts of Asia. Habitats may vary, and evolution may have altered some characteristics of the basic natural identities of blacks. For instance, quantities of melanin substance may remain the same or changed through evolution. A common genetic heritage among the human species exists and cannot be ignored when assessing humans' innate capabilities in capturing, processing, and delivering information among humankind.

Amongst several questions that can be asked with regard to black people's natural dispositions are the following:

a) *If*, for example, melanin granule is responsible for our inadequacies or *if* our behavior is natural and not environmental, doesn't it mean that nothing can be done about it?

b) Most people believe that all things natural are fixed and unchangeable. This general belief was true when studies in natural sciences were at their primary stages (i.e. prior to seventeenth century).

c) The world has changed and is still evolving, especially among nonblack people who are constantly searching for alternatives in everything. They are constantly challenging nature.

*Most* of the nonblack people have proven that they understand the world better than dark-skinned people, considering their rate of progress in all sectors of human life. Black people's stagnation and regressive nature and the nonblack people's continuous hunger to acquire knowledge to improve living standards have caused the world to be divided into three distinct groups: the First World, the Second World, and the Third World. It is a general knowledge that Third World always refers to unprogressive, undeveloped, underdeveloped, or generally, black people's communities and countries.

The Second World tag mostly refers to the so-called developing non-Caucasian (Hispanic and Asian) people and countries.

The First World obviously represents the nonblack race of people—the industrialized, civilized, progressive people. These descriptions are accepted universally by all. Europeans are generally considered to be in category of the First World human beings.

Is this a natural coincidence, natural fact of life, or simply environmental stigmatization?

Some primitive-minded people usually link regressive and progressive differences in nature as spiritually ordained or an act of God. Spirituality as the source of nature has been overridden by science in nonblack communities for centuries.

Since the beginning of the second half of the twentieth century, human knowledge has increased tremendously in the field of natural sciences. Studies in human physiology, the human genome, and neurological sciences, plus the nature of this world have opened several frontiers in human anatomy such that many organs in the human body can now be observed closely or diagnosed with modern microscopic equipment to expose their intricate characteristics while the organic matter is still alive. Organic matters can now be dissected, extracted, replaced, repaired, or medically treated without causing fatal damage to the main body.

Genetic engineering in all organic matter is a specialized field of study at modern universities and general hospitals in most First World countries. New medicines are introduced daily to cure many ailments.

Many diagnostic medical instruments and equipment capable of magnifying invisible details of natural organic components have been invented mostly by nonblacks.

While *most black people* are comfortable with living as usual or maintaining the status quo, *most nonblacks* are constantly asking pertinent questions and finding useful answers.

Most nonblack people are probing deeper into unknown scientific frontiers to make living healthier, happier, and more comfortable.

Unlike most black people who mainly think of surviving today and think far less of tomorrow or the future, most nonblacks think far beyond mere survival. Most nonblacks have dreams for the future; they have long-term plans and are not shortsighted. They dream and see real possibilities ahead of time in areas where most blacks perceive as impossible and therefore leave future possibilities in the hands of their perceived god or divinity.

Yes, if we Negroes can accept that there is something genetically wrong in our genome structure, we can begin to investigate and apply modern scientific techniques that will bring us into equilibrium with the rest of humanity.

Yes, our unique melanocytes produce too much melanin, more than what our brain cells can utilize. The surplus cells mutate or convert into other destructive cells, such as lethal and sublethal genes, to obstruct information processing and delivery. Nonblacks equally possess melanocytes, especially neuromelanin, pheomelanin, and *eumelanin.*

*Nonblacks do not possess excessive concentrations of melanin* in any single body organ, unlike black people. I am certain that melanocytes in black people can be managed scientifically to bring equilibrium.

If indeed this is proven to be our culprit, surely something positive can be done about black persons' natural performances or natural disabilities. Yes, from this perspective, there is a possibility of identifying our natural defect and also correcting it scientifically. I see a real possibility of averting black people's natural mental weaknesses.

The rest of humanity has for centuries mocked the black man for being less of a human being because of black people's unique simplistic nature as caused by certain genes stated above.

E. ASAMOAH-YAW

For example, before 1860, a black person in the United States America was considered as a property, not as a full-fledged human being. A black person was a unit of material asset, such as a chair or a dog owned by someone.

Even the freed slaves did not possess the ordinary basic human title or dignity. He was not a complete being. Basically, he was a thing. He may be freed and out of bondage, but the national constitution did not recognize black people as humans as it did the nonblacks.

The first three people who were assigned by the Continental Congress in 1776 to draft the first US Constitution (Thomas Jefferson, John Adams, and Benjamin Franklin) excluded native people and Negroes in their human population count. Jefferson and Franklin owned several slaves at the time. "Humans" at the time in the Americas were only those full-fledged descendants of the English, Irish, Scottish, French, Germans, and Dutch ancestry.

The difference between a Negro and a sheep was that the Negro was worth more in dollars than a sheep.

He was defined as three-fifths of a human being.

The big question that pops up every time when discussing black and nonblack issues is, why do nonblacks see themselves as naturally better than blacks, or rather, why do some blacks see themselves as naturally naive and gauche? It seems to me that most black people have always viewed Caucasians, in particular, as more sophisticated and knowledgeable or next to God throughout history. And conversely, most nonblacks always view blacks as naturally naive and candid. Can evolution or the environment, rather than nature, make people naive or knowledgeable?

E. ASAMOAH-YAW

Blacks are mostly too simple, less critical, easily convinced, too awkward, and rarely rational.

Some of us do accept this observation as truth. But most of us appear to be adamant about what nonblacks think about blacks. Some of us, the few lucky blacks, especially our intellectuals and academics, always respond to this characterization with rage because they say there is no scientific evidence to support our natural inferiority. The *few* lucky blacks among us, whose intelligence can be measured as equal to that of most nonblacks, are often annoyed to hear such provocative statements from critics.

The only thing wrong with the above assessment is its gross generalization. The statement should rather be *most* black people because *few* naturally gifted black people do exist; possibly, there's only one in every ten thousand who may excel in specific faculties or indeed may excel better than few whites in some areas of human adventures.

Yes, over a long period of time, every unit of every organic matter evolves or changes from their original parent's nature and into something completely different. Yes, this is an observed fact. The stem genes develop into something else for its own survival. Darwinians call this phenomenon random natural selection through adaptation to habitat circumstances. Darwinians are essentially environmentalists.

They claim that over a long period of time— for instance, over a thousand years—all living organisms (including humans) live through gradual organic changes into several distinct divergent genera. Cells will perish if they fail to adapt to environmental forces.

Thus, the humans of today are necessarily not the same as the ancestors who lived about a thousand years ago. We have evolved.

E. ASAMOAH-YAW

It is important therefore to go through history to see if black people's body chemistry contained no melanin, a fewer melanin, more melanin than we have now, or the same as those possessed by nonblacks from the beginning of time.

And if it is established that divergent black people of today really inherited genes of some black people who were once intelligent or even far more intelligent and progressive than every other human race of today, we may then speculate that in several generations to come, maybe a thousand years from now, blacks shall evolve into our past glory of the brightest or possibly even be worse than what we are today. Again the same search can confirm our earlier allegation that nature and nurture together make us who, what, and why we are what we are.

The main purpose of this discourse is to generate curiosity especially among psychiatrists, neurologists, and geneticists to look for genetic sources for black people's inferiority and to apply scientific means to correct it. Yes, I am convinced there is a natural genetic difference or a defect which can be identified and corrected.

And we should stop putting all our failures or blames on the environment. Dozens of research proposals or recommendations, statutory laws, by-laws, and countless new institutions have been introduced for centuries to correct the supposed environmental deficiencies in most black communities, but this melanin-saturated species (the Negro) remains unchanged.

I think it is time the world look elsewhere to properly diagnose and treat the black persons born with this irresponsible difference.

The other questionable organic area which needs to be probed critically is our body's neurons.

Every part of the human body is known to be highly connected with intricate electrical wiring systems, technically called neurons or nerve fibers or cells. Nerve cells handle *information transmission* to and from the brain organ.

There are two types of nerve fibers—axons and dendrites. *Axons* carry signals *away* from the cell body, while *dendrites* carry signals *toward* the body.

The brain organ is the *bank or store* where information is kept.

Nerves are the receiving and delivering information agents of the human body. They serve as messengers or transporters, literarily from the eyes, ears, mouths, tongues, noses, and skin to the brain. After they're processed in the brain, nerves carry the needed information from the brain by ordering the necessary human action. The messengers are encased in myelin sheath, which provides energy needed for efficient and fast delivery of signals or information.

The brain, our organ of thought and center of all nerves, processes information and commands human action.

It is observed in *most blacks* that the end results of this brain-and-body exercise lack quality sensibility or proper judgment when compared with *most nonblacks.* In most black people, it appears the axons and dendrites are naturally deficient and not as efficient as those in nonblacks. In a similar analogy, in electricity power generation and delivery, it is a fact that copper wires are efficient conductors of electric current than lead, iron, steel, aluminum, or any other metallic substance. The implication here is this: black people's axons and dendrites are *likely* to be of low-quality nerves by nature in comparison with nonblacks. The logic in this conclusion is the obvious fact that most blacks are naturally less intuitive, less productive, less inventive, less imaginative, less progressive, less

E. ASAMOAH-YAW

visionary, less adventurous, less virtuous, and less critical.

And the worst part is that most blacks are too comfortable with this mediocrity by nature because we blacks are wired with inferior neurons, which are not capable of receiving and delivering quality signals to and from the brain.

If this conclusion is proven to be the case, as I am sure many will agree to that in the twenty-first century, almost everything is possible. New drugs and new clinical and surgical procedures and techniques are available or indeed can be invented to energize or repair these prevalent genetic disorders in most black people. Genetic engineers can do something by identifying and manipulating the culprit gene.

## The Environment

Conventional wisdom states that all organic matter, including all humans, are what they are mainly because of the environment and their geographical chemistry. There is a universal consensus that black people (popularly called colored people, Negro, African, etc.) are what they are because of habitats or community circumstances. In other words, black persons' being has very little to do with nature, according to environmentalists.

The word *environment* of course embraces everything besides the self.

It includes everything outside the human body. It includes the food we eat, water we drink, air we breathe, sanitary condition we live in, clothes we wear, room we live in, the house we dwell in, the community and the district and the region or area we dwell in, the country we live in, the continent we live in, the world at large, and the climate.

It includes cultures, customs, languages, traditions, laws, institutions, leadership, behaviors, attitudes, procreation, religion, education, and nurturing in general.

Those who claim that people are what they are because of their surroundings do believe that a pregnant woman's living conditions or lifestyle, for example, contributes to the nature of the baby inside the womb; for instance, drugs, bad nutrition, poor sanitation, and unhealthy practices do affect the natural growth process of the fetus in the womb. In short, it is not just the fusion of sperm and egg (organogenesis) and the millions of embryonic changes that take place during the thirty-eight weeks of pregnancy, but it is the nature of these two primary organic elements to stay alive and healthy by adapting to prevailing environmental circumstances of the pregnant woman. Organic adjustment to the environment is inevitable.

Because of these and many other social factors, we cannot say that nature is the *only* designer of the self.

All the above environmental factors are true.

Indeed, human life would be meaningless without most of the factors stated above. My emphasis, however, is that nature or the human organic matter comes first before the environment. The autonomous organic matter with its complex natural design contains its unique natural property, its structure and form, which the *environment cannot redesign*—at least not yet.

It is up to that organic matter to tune or adapt to the environment for its own survival. The organic matter's nature definitely comes first.

Question is, if blacks were as smart as whites, why didn't black people enslave or colonize white people in Europe, for example? It is known that blacks practiced slavery among themselves in

Africa for centuries before Europeans emerged in Africa.

There is historical evidence that black people in the west coast of Africa were offering war captives (blacks) to European traders in exchange for manufactured products (such as beads, mirrors, machetes, alcoholic drinks, and clothing) centuries before institutional slavery commenced.

There is historical evidence that buying and selling of fellow humans prevailed in all racial groups (including Europeans and Asian people) and not just among blacks only.

Human being as a unit of commodity for sale has been with mankind since ancient times. There is no culture in the world that has never indulged in slavery in its entire history. Sale of fellow human beings to other humans was common in all racial groups. I will keep repeating it!

Among the *popular reasons why enslavement of blacks* has been going on for so many years is the following: blacks are naturally peaceful, naturally hospitable, naturally simple, and naturally respectful, and for these reasons, blacks do not possess an instinct to detect slave masters' wicked intentions of how the sold slaves would be treated in confinement by those who purchased them. Every kingdom in black Africa had indulged in selling human beings (war captives) for centuries before the white people entered Africa.

Slavers enslaved blacks because they saw black people's mental weakness and knew they could use black people as tools with negligible challenge (i.e. like a machete, a horse, a tamed animal, a tool, or a household equipment) and without problems.

They knew most black people could be dehumanized because most blacks demonstrated

E. ASAMOAH-YAW

then (and still exhibit) pettiness in everything apart from entertainment, sports, and mimicry. *Most* black people have historically performed very poorly in creativity.

Most black people are the most impoverished race of people for millions of years, yet black people are known to occupy the richest portions of landmarks in the world. Ironically, blacks are potentially rich, yet they remain the poorest and the most wretched among human kind. Why shouldn't we blame nature for being what or who we are?

Why should we continue to blame the environment when it is empirically proven that every environmentally proposed solution has been tried and has failed in all black communities?

There are two countries whose founding histories can be used as evidence of needlessness in linking black people's mental weaknesses with the environment: Liberia and Australia.

They passed through similar historical circumstances. The two countries were formed by social deviants and dehumanized people: British ex-convicts for Australia and ex-slaves from the Americas for Liberia.

Both countries possess numerous natural resources with similar climatic conditions.

The two nations were previously occupied by several indigenous tribal people before the new foreign migrants were brought in to settle in the area.

Liberia was founded by free American and Caribbean slaves, while Australia was founded by prisoners, convicted criminals or felons in British prisons who were either purchased with cash or offered free to shipowners and exported as human cargo to work on plantations in

Australia. Both countries have been managed by expatriates since their establishment.

Australia's population has been essentially white people and governed by nonblack people. Liberia's population is black and governed by black people. The land quality in both countries is similar—potentially rich with abundant mineral resources. Australia (white people's country) is today one of the richest countries in the world, while Liberia (black people's country) today is one of the poorest countries in the world.

# CHAPTER 7

# Black and Nonblack Countries Compared

## Liberia

THE AREA SIZE of Liberia is 43,000 square miles. Population is 3 million. By the year 1820, the black population in the United States of America had reached its optimum level. Shiploads of more black African slaves were brought to the Americas every day for sale. The whites foresaw future potential threat to their racial domination and supremacy. The black population was increasing at a faster rate than white people; hence, measures should be taken to prevent a repeat of Moses's story in the book of Exodus in the Holy Bible.

White people bought black people to work on plantations in the Southern states and partly to breed the blacks for resale purposes. Indentured male and female black servants were forced to have as many children as required by their slave masters. Breeding of Negroes was a secure investment.

The less useful blacks were publicly auctioned on public platforms or exchanged for land or property or cash.

Struggle for freedom among the slaves had increased in most communities because of the prevailing inhumane practices. Runaway slaves and slaves killing slave masters were very common, and slavery as an established, legitimate institution was cracking at its base. White people realized that most black people and mulattoes would prefer to return to their African motherland than to live in servitude.

Thus, in 1821 the American Colonization Society bought a large piece of land (43,000 square miles, almost half the entire new country) in the west coast of Africa, Cape Mesurado.

The site then was called Grain Coast by the Portuguese because of its valuable crop called pepper. Different European traders called the area by different names, including Pepper Coast, Grain Coast, and Malaguetta Coast. The initial settlement was funded by free American and Caribbean societies.

In 1822 the first freed African American slaves with mulattoes were shipped to form the Monrovia settlement. The first site of settlement was named Freetown. A larger settlement area was created and named after the fourth US president, James Monroe (1817–25). Monrovia has always been Liberia's capital city to date.

In 1847 Liberia became an independent republic with a black president named Joseph Roberts.

At the year of independence, total population was about three thousand men, women, and children of American African slave descent transported back permanently to the African soil.

There were sixteen distinct tribal groups already living there as their natural habitat.

The American slave masters at the time thought they were returning their African slaves back to their natural homeland in Africa. The only criterion was their skin color—black or half-caste people. They did not realize at the time that these were descendants of over 12 million African slaves who had been out of Africa, their motherland, and had been dehumanized for two centuries and therefore had no idea of where or what the continent of Africa was or looked like.

It was impossible to imagine or to trace their original ancestors' roots or their natural cultural heritage or tribe. These people were neither Africans nor Americans.

They were descendants of African Americans who have been given the right to manage their own affairs in their own original natural habitat. A hatched bird's egg in a cage when matured cannot behave normally as its kind in the wild when set free.

After over a hundred years of settlement in Liberia, these people could not manage their own affairs into progress and prosperity. They could not adapt to the norms of the sixteen original indigenous tribal people. Apart from their natural resemblance, the returned freed slaves and the local people were completely different from each other. One had to coerce the other to achieve cohesion.

Thus, the new nation was saddled with socioeconomic and political problems from the start. It was taken for granted that over one hundred years of conversion from "savagery" to "civilization," the new nation, Liberia, would become autonomous and progressive. This was self-deception. The new country could not utilize the abundant local natural resources to establish themselves as true human beings who are capable of ingenuity and progress. They exhibited the usual nature of the black man— very comfortable with mediocrity and lack of creativity.

By Liberia's example, it can be concluded that blacks in general have been good at nothing good for centuries no matter the type of environment blacks have lived in. The slave masters knew that the black man was (and still is) naturally incapable of managing his own affairs no matter how fine you nurtured him.

Liberia received several substantial loans to pay off its years of accumulated foreign debts from the US Treasury and private sources, including Firestone Tire and Rubber Company of USA. This company had secured 400,000 hectares of land for a rubber plantation in Liberia. The company and its subsidiaries, plus other new foreign ventures, did well initially to induce change and economic growth when the whites were managing their own foreign investments in the country. But few years later, when black people took full responsibility in the companies' management, the companies and the country at large began to deteriorate just like the rest of black African countries south of the Sahara.

Thus, the country's economic declining process, which began at day one of its foundation, accelerated at full speed with multiple social problems, corruption, decadence, military coup d'état, new constitutions, and civil wars claiming over 150,000 human lives all because of the black man's unique inferior natural makeup. Liberia, the oldest republic in black Africa, is still (in the year 2016) among the poorest nations in the world; 68.6 percent (2.42 million people) out of the total population of 4.4 million live below the poverty line of $1.90 daily (April 2016, World Bank statistics).

These two countries, *Liberia* and *Australia*, have had a similar historical experience worthy of comparison between black people and nonblack people's natural disposition.

## Australia

Australia is the smallest continent among the seven on planet earth.

- area size: 2,9 million square miles
- population: 23.5 million (2011)

- ethnic groups: 95 percent whites, 2.5 percent native Australians, 1.3 percent Asians, 1.2 percent others (2011 population count).

Note: The official name of the country is *Commonwealth of Australia.* It includes the Australian continent, Tasmania island, and several other smaller islands.

Prior to the year 1788, the continent of Australia was occupied by over two hundred different ethnic groups of dark-skinned people.

Britain at the time consisted of two distinct cultures (England and Wales). The Act of Union of 1707 brought the English-speaking and Welsh-speaking (Celtic) people together to form Great Britain.

The Act of Union of 1801 abolished the Irish Parliament and added Northern Ireland and Scotland onto Great Britain to form the United

Kingdom. The number of people who had been convicted as criminals by courts in Britain and, later, the United Kingdom had increased to an unmanageable level such that some prisoners had to be sold or given freely to (British) plantation owners in Australia and the Americas.

Although Australia at the time had been visited earlier by the Dutch and other European adventurers who were searching for products which were either scarce in their countries or new products which were not available in Europe, it was the British who established Australia as a stable colony with hardcore criminals from British prisons.

Several shiploads of convicted felons were exported from British ports to Sydney, Melbourne, and other ports in Australia between 1770 and 1850. These were the early settlers or the founding fathers of Australia. *Free settlers* arrived in Australia not earlier than the midnineteenth century.

It is important to note that the land was occupied mostly by native black people prior to the British white settlers.

By the end of nineteenth century, the black indigenous population in Australia was almost extinct. In order to have full control of the land, the white settlers had virtually exterminated all the blacks.

Survival of the fittest is a natural phenomenon in the world of organic matter. World history from prehistoric times has plenty of examples which are similar to the Australian experience. Those who possess inbuilt natural capabilities for self-defense always remain alive and continue to progress. Blacks have historically shown natural mental deficiencies, and nonblacks recognize these and take advantage of them in the same way as humans see animals' natural mental deficiencies and take advantage of animals.

Those who lack this inbuilt quality always pay heavily. This is not by any means a justification of the brutal acts of white people against blacks or the Australian Aborigine population. It is a statement of fact that there exist natural similarities, natural differences, and natural equalities. We are all *Homo sapiens*, but some are more human than others. Such is the painful fact of life.

While the British were losing their colonies in the Americas, they were advancing their stronghold in Asia, Africa, and particularly, the Australian peninsula. The declaration of unilateral independence in 1776 by European immigrants in North America intensified the British move to colonize Australia by any means possible.

As a colony, Australia was managed initially by merchants and later the British Parliament as indeed was the case in all British colonies in Africa. At that time, all the colonies were booming with progressive measures toward economic growth and prosperity because the leaders at the time were humans whose biochemistry contained quality cells which have the ability to perceive progressive ideas and enforce rules and regulations for the benefit of the Caucasian populace.

Leaders in government were pursuing public interests, while those in industry were destined for private sector interests.

Both public and private sectors harmoniously performed their best for the benefit of the new nation.

Most importantly, all these leaders were not black people.

E. ASAMOAH-YAW

They were white people (Caucasians) who by nature were different human species with *very low* amount of melanin in their neurons or body chemistry. This assertion has been repeated several times in this discourse because it appears to be the major natural difference between blacks and nonblacks.

Australia has continuously progressed, and indeed, it is one of the leading industrial countries in the (2016) world. Surely, its accomplishment is not by means of luck or an act of God as religious addicts may conclude. It is the result of quality brainwork and intellectual performance which ironically had been inherited from British social deviants of the past.

One would naturally have expected such class of people to be low performers, but because of their natural biological heritage (high-quality neurons), their performance is comparable to that of the people in their motherland.

It is predictable that if all white people of Australia were to move to Liberia and all blacks of Liberia also were to move to Australia in exchange today and allow each group to perform with the same unrestricted conditions for five years, the whites in their new settlement would show significant progress in every sector of the country, while the blacks would show lower than significant progress or possibly show negative or regressive progress. It's because black people for several centuries have demonstrated natural disability in managing their own affairs without intellectual or technological support from the white man.

## Congo Free State

The Democratic Republic of Congo—this country was initially bought and owned by one person between 1884 and 1908. It is a classic portrait of the natural differences between black and white.

The Congo Free State was a huge territory with an area size of about 905,000 square miles in west central Africa that *was the world's only colony claimed and owned* by one person between 1844 and 1908.

That man was called King Leopold II of Belgium.

The king was born in 1835 and died in 1909. He inherited his father's throne as the king of Belgium in 1865 till death in 1909. His official status as a king began in 1864 when he was twenty-nine years old. As a young man and a son of the former king, Leopold II was indeed far more privileged than all his contemporary leaders. His heritage was intertwined with most European kingdoms through intermarriages which had linked him with every leader who mattered at the time in Europe. His brothers, sisters, nephews, uncles, and cousins were the aristocrats of most European kingdoms.

He knew from infancy that earlier powerful Europeans had conquered and established huge empires or territories across the globe, particularly the continent of Africa.

His kingdom, Belgium, had no colony at the time. And instead of looking for a colony for his country, he dreamed of possessing one for himself so he could boast of being the single owner of a massive land with all its contents (animals, trees, rivers, humans); it would belong to Mr. Leopold himself and not for Belgium State. His focus was on the continent of Africa because most European powers at the time possessed several colonies there at the time. Africa was a continent of savages, destined to be grabbed and civilized by Europeans.

In nineteenth century, the entire African continent was seen as a huge mass of land occupied by savages, barbarians, and uncivilized human beings who lived on trees, waiting to be Europeanized. One person who was popular in most popular newspapers at the time was an Englishman described by all as Henry Morton Stanley, the explorer.

By 1850, the continent of Africa consisted of 700 different countries at least or comprised of a distinct group of people who occupied defined geographical areas characterized by uniform cultures.

No Black African country then had originated literate works of any type. None envisioned a need for literacy.

Each African country had its language; organizational systems; legal systems; religious systems; and economical, social, and political systems which had been practiced for centuries. By the end of 1950, these countries had reduced to about 50 confused nations—not by choice of Africans but by the coercive influence of Europeans without the slightest consultation of any one of the 700 countries or leaders in Africa.

The question is, *if* Africans were as intelligent as Europeans at the time, or since then, it certainly would have been impossible for such ordinary white people like Henry Stanley to cause so much incurable ailments among the black populace on African soil. Whatever their method or the motive, the Africans at the time must have outwit them with sheer common sense and intellect. The blacks could not because they were not smart. The whites were (and still are) smart.

Let us examine the Congo affairs.

Facts to remember: King Leopold II of Belgium in person never set a foot on African soil nor ever saw a drop of African blood spilled by Europeans. Sir Henry Morton Stanley was born in Wales in 1841and grew up in an orphanage home in Britain.

He migrated to the USA and became an American citizen at age sixteen. He served in the US Confederate Army during the Civil War. He became a journalist and explorer of Africa. The following is a brief record of Stanley and the king of Belgium, who employed him to go to Africa to purchase land which the king named as the Congo Free State.

1. HENRY M. STANLY and his white assistants had used a variety of tricks such as fooling Africans into thinking that whites had supernatural powers, to get Congo chiefs to sign their land over to Leopold. For example: a number of electric batteries had been purchased in London, and when attached to arm under the coat, communicated with a band of

ribbon which passed over the palm of white brother's hand, and when he gave the black brother a cordial grasp of the hand the black brother was surprised to find his white brother so strong, that he nearly knocked him off his feet. When the native inquired about the disparity of strength between himself and his white brother, he was told that the white man could pull up trees and perform the most prodigious feats of strength. Another trick was to use a magnifying glass to light a cigar, after which the white man explained his intimate relation to the sun, and declared that if he were to request him to burn black brother's village it would be done. In another ruse, a white man would ostentatiously load a gun but covertly slip the bullet up his sleeve. He would then hand the gun to a black chief,

E. ASAMOAH-YAW

step off a distance, and ask black chief to take aim and shoot; the white man unharmed, would bend over and retrieve the bullet from his shoe. By such means, and a few boxes of gin, the whole village had been sold away to your Majesty. Land purchased in this way, Williams wrote, was territory to which your Majesty has no legal claim, than I have to be the Commander-in-Chief of the Belgian army.

2. In another instance, Stanly is well remembered of his broken promises, his copious profanity, his hot temper, his heavy blows on the Africans, his severe rigorous measures, by which the Africans were evicted of their lands because the land had been purchased by King Leopold. This is a personal experience narrated and recorded by several eye witnesses

such as George Washington Williams, William Sheppard, Edmund Dene Morel, Joseph Conrad, and Casement. William's dissatisfaction of maltreatment of native people circulated widely at the time. Here is a few of it.

3.  His Majesty's Government is excessively cruel to its prisoner, condemning them, for the slightest offences, to the chain gang as often these chains eat into the necks of the prisoners and produce sores about which the flies circle, aggravating the running wounds. In the meanwhile, King Leopold was falsely claiming that his personal new state was providing wise government with good services to the natives. But there was evidence that there were no hospitals, no schools, except for few sheds not fit to be occupied by a horse. Virtually none of the officials working in the State spoke any

E. ASAMOAH-YAW

of the African languages. Reports of His Majesty's Courts were abortive, unjust, partial, and delinquent. For example a white servant of the Governor General went unpunished for stealing wine, while black servants were falsely accused and severely whipped. White traders and state officials were kidnapping African women and using them anyhow, few lucky ones were kept as concubines. White officers were shooting villagers, sometimes to capture their women, sometimes to intimidate the survivors into working as forced laborers and sometimes for sports. Two Belgium Army officers saw from the deck of their steamer a native in a canoe some distance away. The officers made a wager of five pounds that they could hit the native with their rifles, three shots were fired and the native fell dead. The bullets pierced through the head. In the 1890s slave trading was abolished, especially in the Americas, but Leopold's Congo was selling humans for three pounds per able-bodied person to shipping companies destined to the Americas. Williams reported that His Majesty's Government in the Upper River was composed of slaves of all ages and both sexes.

In Williams's open letters (one was to the then US president Harrison), among many complaints, he wrote that at Stanley Falls slaves were offered to him in broad daylight and that at night he discovered canoes loaded with slaves bound strongly together. He pleaded that the oppressive and cruel government of Congo Free State be replaced with a new regime that would be local, not European; international, not national; just, not cruel. There were several other reports of outrageous inhuman activities from Casement, E. D. Morel, and Joseph Conrad.

To legitimize his position as the sole owner of Congo Free State, Leopold organized a conference (the 1884–5 Berlin Conference) where he invited all European colonial powers, United States of America, and Turkey and declared the continent of Africa (the Congo Free State) as his personal property. He further asked the other stakeholders to clarify their African holdings around Niger River and Congo River.

Each power marked its territory with chosen colors on an African map where visually territorial boundaries could easily be identified. This conference affirmed the *British* claim of Egypt, Gold Coast, Sudan, Gambia, Kenya, Uganda, Rhodesia, and Nigeria. *France, Spain, Portugal,* and other powers declared their colonies at the conference with different colors, but actual formal ratification took place during the ensuing years.

The king was proud among his peers that he had successfully done what no other leader had ever dreamed of in Europe.

International pressure from his peers in Europe and USA, of the deadly practices in his rubber plantations, ivory trading, and gross abuses of human rights made the king cave in, and he handed his property, the Congo Free State, to the government of Belgium in 1908. During that period, historians estimate that millions of blacks lost their lives, and those who survived were disoriented and never recovered.

King Leopold II of Belgium and several European companies at the time were able to amass wealth unprecedented in European history for a short time—only a thirty-year period. This is the result of defenselessness when two species of unequal brain power meet face-to-face. The best naturally wins always.

Some historians conclude that it was sheer wickedness of some Europeans of the period or natural mental weakness of most blacks at the time and place. Well, since this is historically not an isolated case and has been recurring throughout ages, the blame fingers should be pointed at the victim, the black man, for being perpetually doddering when in contact with nonblacks. I am here therefore to appeal to all people without prejudice to critically think about these centuries-old differences between blacks and nonblacks and *prove me wrong* that racial intellectual difference is not natural or not in the gene but in the environment. It is not enough to just say that scientific evidence doesn't exist and that necessarily all humans are equal.

All humans are not the same; even identical twins are never the same.

Why can't scientists investigate deeper into human genome and neurons very closely this time? I am convinced that the evidence of my proposition—*that* most *blacks are naturally less intelligent,* few *blacks are smart,* most *nonblacks are naturally smart, and* few *nonblacks are naturally not intelligent*—is hiding in these two areas of human chemistry. No one can deny the prevalence of human natural similarities, differences, and equalities.

Why is humanity in denial of acknowledging several centuries of human earthly experiences of most blacks' ineptitude when compared to most nonblacks?

# CHAPTER 8

# Racial Performances Compared

THE HISTORIES OF Liberia, the Commonwealth of Australia, and the Congo Free State, as briefly written above, present classic examples of the natural intelligence differences between the black people and the nonblack people. The three country's circumstances had virtually nothing to do with their environments.

A summary of World Bank Data of 2014 shows Australia had the world's fifth largest per capital income. It has the second largest developed index globally. Australia ranks high in many international comparisons of national performance, such as quality of life, health, education, economics, freedom, protection of civil liberties, and political rights. It is a member of the G20. Australia was a founding member of the United Nations. In terms of average wealth, Australia is ranked second in the world after Switzerland in 2013.

Above all, Australia is a developed country and is one of the wealthiest in the world, ranking as the twelfth largest economy.

There is no doubt that environmental factors also contributed to Australia's unparalleled success, but the deep-down reason is the content or genetic makeup of the persons who made this success possible. The leaders possessed natural capabilities of organizing themselves, making useful laws, enforcing the laws in such a hostile environment (descendants of criminals) and on a people who were known as disobedient to the laws in their original motherland. The nonblack settlers have not only managed the available resources to satisfy themselves but were also able to manufacture goods and sell beyond their borders. They are progressive and inventive. They are capable of identifying their current needs and that of their future, and they do everything possible to accomplish them.

For barely two hundred years, Australians have demonstrated that with superior brain power, nothing is impossible. These successes have been possible because of their natural makeup, plus the nonblacks' natural inclination to challenge nature. Their body chemistry naturally contains very low and insignificant amounts of melanin, especially inside the brain organ. Black people's brain contains more melanin in every part of it when compared to those of nonblacks, especially those of Caucasians.

When we compare Liberia's history and circumstances to that of Australia to date, we see that the growth levels of both countries are completely different. While Australia shows consistent progress in all sectors, Liberia is consistently in decline.

According to World Bank Data published in April 2016, 68 percent of Liberians, the first African Republic, live on the poverty level ($1.90 a day). In Ghana, 25 percent of the populace lives below $1.90 daily.

All black African countries have high levels of poverty and also do exist today because of developed countries' willingness to offer them financial and technical support.

Black Africans have lived in their environments for centuries and still depend on nonblacks for everything. Black Africans are naturally "blessed" with every conceivable natural resource, yet not a single African country can survive a day *if* all technological gadgets and nonblack support are withdrawn from them.

E. ASAMOAH-YAW

Blacks are stagnant and unproductive because of mental weakness. It is our behavior which motivates whites to take advantage of us. We have always served as bondsmen or chattels to nonblacks because we always exhibit a sense of vulnerability to them. We are mostly ignorant of our nature. We blacks need to understand this as our nature and also do something about it. It is possible in the year 2016 that science can correct our natural defects.

Liberia is therefore a classic example of black people's natural performance weaknesses; that is, when given the same chance as the Australian nonblacks had, Liberians could have performed similarly or even better than the Australians if the two races possessed identical biochemical characteristics.

Founders of both nations were set free with the expectation that they could manage themselves better, but it has been proven that all humans are *naturally not* the same. We are equal and similar by appearances but naturally different by content.

Australians and Liberians from the beginning had equal chances to decide their destinies, to prove their best, to organize themselves with good institutions and dedicated leaders, to make good laws and enforce the laws, and to be creative, inventive, and progressive. Liberians, like their neighboring West African colonies, recognized *only* their freedom from slave masters' oppression and suppression, hence their indulgence in leisure and pleasure instead of conceptualizing effective ideas for self-improvement.

Living in what was formerly known as the Grain Coast, they saw no need to cultivate or improve the land use for grain plantation or to exploit the abundant natural resources. Instead of building factories and manufacturing for themselves, they were building churches and importing manufactured goods from those who oppressed them. Most of black people's neurons always fail to capture and process information critically, therefore leaving the black person helpless, hopeless, aimless, and unproductive.

In all sincerity, the blacks in Liberia and the whites in Australia have shown the world vividly the natural performance difference between the two human species when given equal opportunities to manage their lives.

Records of the past independent states such as Liberia in 1844 and Ghana in 1957, followed by the independence of over fifteen black countries in 1960 alone, would have shown that something good about the Negroes would come out to show the world that black man is actually 'capable of managing his own affairs' as Kwame Nkrumah (Ghana's first president) proudly declared on independence day (6 March 1957).

Black men are still living in the past and somehow appear to be begging their former colonial masters to come back to govern them. There is something internally wrong with black persons by nature. It can be traced from the internal organs, especially the organ of thought or the neurons. These can be manipulated medically, clinically, or surgically to normalize the blacks' nature.

Yes, it is definite that in the year 2016, most human organs can be manipulated scientifically successfully.

By the year 2000, not a single black country could have survived independently without nonblacks' intellectual and financial support. If this cannot be a conclusive evidence of black people's natural mental weakness, then I don't know what else to blame.

E. ASAMOAH-YAW

# Negroes and Caucasians

Here's a brief history of Caucasians' encounter with West African blacks. All kingdoms in all continents had been at war with one of few of their neighboring cultures sometime in history. A feeling of insecurity among dynasties, kingdoms, and nations was the experience of the past. Old kingdoms were overthrown, and new ones emerged.

In Europe, Portugal was the first nation to attempt to expand its influence beyond continental Europe, starting from the fifteenth century. Although most European nations had curiosity about the nature of the Dark Continent, they were not certain. The Portuguese's initial motive was acquisition of knowledge or exploration. The second reason was to be the first to find a sea route to India. Stories of Asiatic kingdoms such as India, China, the Mongols, and the rest were fascinating topics at the time. The Portuguese's second reason was to bring Christianity to Africa and, if possible, to convert Africans into Christianity.

Trading with the native people was another motive though not necessarily the major one at the time; broadly then, the Portuguese entry into Africa was (1) scientific, (2) religious, (3) economic, and (4) political.

Under the inspiration of the Portuguese prince Henry the Navigator (1394–1460), who initiated experiments in shipbuilding and navigational equipment and techniques. For the first time, sea vessels could sail on deep oceans to places like India. In the year 1420, he began to send ships to explore the Atlantic coast of Morocco and, by 1444, Senegal. Before his death in 1460, Prince Henry's ships had reached the coast of Sierra Leone. He later dispatched fifteen improved ships for exploration. Portuguese ships had reached Benin in 1475, the mouth of River

Congo in 1482, the Cape of Good Hope in 1487, and India just before the end of the fifteen century. It is further recorded that a small party of Negroes were taken from Senegambia area to Prince Henry in Lisbon, where they were trained to become Christian missionaries.

Thus, the Portuguese's initial adventures into African coasts, as stated above, marked a significant historical turning point of the imperfect relationship between blacks and nonblacks. The *Portuguese* pioneers had four colonies in Africa before 1920. They were Angola, Guinea Bissau, Mozambique, and Sao Tome Principe.

The Dutch followed in the seventeenth century and ended with two colonies of Democratic Republic of Congo and Congo. France grabbed twenty-two colonies. These were Algeria, Benin, Burkina Faso, Burundi, Cameroon, Central African Republic, Chad, Comoro, Djibouti, Gabon, Guinea, Ivory Coast,

Madagascar, Mali, Morocco, Mauritania, Niger, Rwanda, Senegal, Togo, and Tunisia.

The English had acquired seventeen territories. These were Botswana, Egypt, Gambia, Ghana, Kenya, Malawi, Namibia, Nigeria, Seychelles, Sierra Leone, Somalia, South Africa, Sudan, Swaziland, Tanzania, Uganda, Zambia, and Zimbabwe. The Spanish and others' earlier interest in Africa were marginalized by the then superpowers.

Thus, until the beginning of the twentieth century, Caucasians and blacks' intercourse of the past throws vivid light on to the natural difference between the two races. The relationship was one of master and servant.

With his ideas of seeking adventure or probing beyond his coastal lines and actually designing and constructing ships with metals that can sail on deep seas without sinking, plus the idea of designing navigational equipment that can enhance accurate naval travels, Prince Henry was indeed genius.

Again after close encounter with black people for the first time, he studied them and found that he could use them. He was the one who opened the gateway between Europe and the rest of the world and, in particular, the colonization of Africa.

The prince's imagination could have come from anyone among other Europeans eventually but certainly not from a black African because blacks have shown for centuries that they are naturally not creative and progressive. This explains the reasons the Portuguese were the first to build castles and forts along African coastlines to explore and exploit virgin lands of the Dark Continent.

*If* the black Africans in those coastal areas were as intelligent as the white intruders, they would have used their own manufactured weapons to scare, to fight, or to prevent those foreigners from encroaching into local affairs.

There is no way the Portuguese and other European nationals would have lasted as long as they did if the blacks were as smart as the whites (i.e. possessed much powerful weapons and manufactured goods locally). Europeans successfully used these inventions to entice the black leadership. The Europeans entered African coasts with the clear intention of tricking, cheating, stealing, intimidating, and converting the black Africans from so-called barbarism to civilization.

They successfully attempted all the above and finally made the Africans believe that the white men were far more superior human beings than them. The site of a steamship with heavy equipment on the surface of the sea without sinking was miraculous to the blacks because they could not imagine such possibility.

E. ASAMOAH-YAW

Throughout 6,000 years of recorded human history, the pure-blooded *Negro* has invented virtually nothing significant. I am aware that sometimes, *once* in a generation, a black person may invent something new, but when compared to nonblacks' inventions, our contributions are negligible. There are thousands of evidence everywhere in *every* generation among them. Just look around yourself and count how many gadgets or things in the immediate surroundings are invented or improved by people of color.

Several new products are popping up every day in every generation among nonblacks. I like to repeat here that this brainwork is not an act of divine powers because verifiable evidence does not exist to support the divine element.

There is a definite, supportable natural difference between black people and nonblack people.

# Anthropology

An African dentist can tell a Negro's tooth from a Caucasian at a glance.

Negroes have arms which are longer, relative to body height, than those of Caucasians. This feature, together with their much thicker cranial bones, gives Negro athletes an advantage over Caucasians in boxing.

Another interesting phenomenon is the black person's contempt of the Negroes' image when they progress from wretchedness to affluence. Most of the time, when black people manage to rise from rags to riches, they see themselves as part of the affluent white folks. Their natural image from top to toes must be transformed to resemble nonblacks purposely to show the world how repulsive blackness is.

First, they bleach their skin to look light. Second, they straighten their hair from curly and woolly to stretched blonde floating over their shoulders. The inherited thick and protruding lips, the flat recessive nose, the chin, and the cheekbones would be surgically transformed, proudly, into copycat counterfeit Caucasian. They are proud to be rich and famous but ashamed to live with this melanin-loaded body.

My concern in this discourse has very little to do with the Negro complexion being transformed with the aid of cosmetics. My focal point is on the internal organs, which control black people's behavior. Cosmetology gives a false impression and false hopes. It does not alter body neurons, the brain organ, and its components, especially the excessive melanin content in such a black person; it still remains constant in the body chemistry.

# CHAPTER 9

# Negroes' Leadership

MANY SOCIAL SCIENTISTS have come out with an idea that any organization that has good leadership can achieve all conceivable goals irrespective of the organization's racial environment or the leadership's racial background.

Hence, being normal human beings, black people can be as good as any racial group if they are managed by good leaders of their own kind. According to experts, what matters most is the leader's perception of *good leadership* and its inherent responsibilities. The black or nonblack nature of that person or that organization is considered irrelevant.

*Leadership* as a word comes from *lead,* an old English word which represents *top, tip, chief, principal, main, central,* and *prime.* Anything that leads is called leader, a frontrunner or a spearhead. Human beings are known to be political animals. We have always lived in organized groups of families, friendships, households, villages, townships, and nations linguistically, traditionally, customarily, and religiously with the intent of resolving common social problems together.

This is a universal human behavior. For this reason, a hierarchical arrangement of the total population into groups and group responsibilities has always been prerequisite of every human organization. Thus, on a pyramidal arrangement, one person prevails to present leadership for the total population.

Again, in all human organizations, a few other people are selected to deputize or assist in managing or running affairs for the community's common good.

Leadership criteria and qualification vary from culture to culture and generation to generation. In the past, leaders prevailed through gallantry, physical force, mental persuasion, hereditary rights, and more often, aggression by any means possible or domination of the weakest by the fittest. And an ideal leader therefore is one who has the natural ability to manage all affairs of the organization to achieve safety, health, cohesion, and economic and political progress.

In short, a leader must be prudent, productive, powerful, assertive, and above all, diplomatic or sensitive to every public issue.

According to one ancient philosopher, Plato, a leader must be a philosopher-king, a person born with innate qualities to organize, to direct, to control, to supervise, and to plan the affairs of an institution. A leader manages resources with the intent of accomplishing progressive goals. Leadership is a gift of nature and not environmentally acquired, according to Plato.

A glance through African history from ancient times to date will show that several kingdoms and dynasties thrived for centuries in their respective regions. And certainly they could not have survived for so long if they were not run by good leaders.

For instance, the ancient Songhai Empire, which existed between AD 500 and AD 1200, covered the areas below Sahara Desert or the Savanna belt between Senegal River and the bend of Niger River. This empire disintegrated into two major empires of ancient Ghana and ancient Fulani then again redeveloped into the autonomous kingdoms of Mali, Wangala, Asante, Gonja, Mossi, Hausa, Oyo, and several other powerful states in the forest regions of West Africa. On the central and southern regions of

Africa also, several kingdoms under powerful leadership excelled and for centuries passed through the inevitable ups and downs a nation experiences akin to every human organization.

Yes, such leaders as Sunni Ali, Mohammed Askia, Mansa Musa of Songhai Empire, Osei Tutu of the Asante Kingdom, and evidently many dozen others existed before. But can we honestly measure their legacies in terms of monumental edifices and mentors they left behind? The answer is yes and no. They died and left behind virtually nothing—no solid foundation for their successors or future generations to build progressive societies on. There were only a few exceptions who can be praised for leaving behind useful institutional structures; some of which are still existing.

Truthfully, they acquired neighboring lands and enslaved their inhabitants by force, but they failed to energize the populace to be politically and economically self-sufficient and imaginative.

*Leadership* in black people's environment appears to be an alien word. In all social organizations, there is always a front person, a spokesman, or a spearhead responsible for planning, organizing, direction, supervision, and control of the people's behavior toward general livelihood improvements.

But because of black people's unique nature, their societies have not, for several centuries, had any leader who has successfully exhibited the above five qualities of an ideal leader. Today's world of the Negroes is the picture or the full story of their ancestors' performances.

Throughout black leaders' tenure in office, they always appear to be engaged in infantile experiments, shoddy public projects, and favoritism toward their financial supporters and cronies. Most often, they behave like puppets of a political party's leadership and directives, most often grossly dependent on nonblack leaders' support. In situations where a blind person leads blind people, there is always disaster, stagnation, complacency, and anarchy. And I firmly believe that if Negroes' nature were the same as nonblacks, there would be at least one or two black leaders in human history whose posterity one could boast of, one who has changed the mind-set of the people being led, into universal, long-lasting prosperity.

It is very tempting to cite one or two African leaders of the nineteenth and twentieth centuries as classic examples of ideal leaders in Africa, but how long were these allowed by their people to perform their best? Or how could these virtuous leaders not prevent their enemies from overthrowing them? For instance, *in collaboration with nonblacks*, the military leaders of the first president of the Republic of Ghana (1960–66), Kwame Nkrumah, overthrew him.

His social, political, and economic programs, which were intended to change traditional mind-set of the populace, were not allowed to materialize. Blacks are historically mostly temperamental, impatient, and looking for quick solutions to everything that requires critical thinking. This is obviously a typical example of black people's naturally weak mentality.

Most black people's behavior cannot be anything other than our nature. There is definitely a missing link with humanity.

Blacks are mostly always gullible and amenable to whims of nonblacks, especially Caucasians, due to the blacks' natural faulty neurons. Yes, prior to the nineteenth century, black Africa did have a few leaders who attempted to cause permanent changes in their traditional areas, but their concerns were primarily expansion or acquisition of new lands and shortsighted control of the inhabitants. Their intentions were not for building monuments or mentoring future leaders. They chose to be charismatic leaders to be worshiped by ignorant citizenry.

They failed to introduce literacy and new norms to replace unproductive old customs. The people's mind-set remained constant and unimaginative as it was centuries ago.

African leaders of the twentieth and twenty-first centuries appear to be somehow different. Their primary aims of becoming leaders is not for national territorial expansion, as their predecessors, but they're rather motivated by personal or private interest—stealing public money and saving them in nonblack countries, buying big cars and big houses, and generally indulging in affluent living.

Occasionally, a politician emerges with public ambition. But in the African context, these occasional leaders aim at building personality images under the pretext of building monuments, such as schools, hospitals, and roads with the hope that they would be remembered in the future of the fragile structures they leave behind.

Unfortunately, these public projects or monuments are so badly constructed that most of them do not last beyond three years. These supposed modern black leaders are not fit to be called ideal spearheads. As already mentioned, a leader who will not fail to generate or nurture a generation of mentors who can practice good leadership toward continuous general progress or leadership which will leave a solid foundation upon which posterity can maintain and promote progress affecting every aspect of human life has not as yet been born among blacks.

Modern African political leaders are individuals who have been selected by political parties to represent the party functionaries' private interests. They occupy constitutionally declared public positions as the president, the minister, the deputies, and several others with clearly defined responsibilities, but when in power, the public interest changes into interest of the self, family members, cronies, and colleagues who financed them into their current public status. And without hesitation, it is unlikely that ideal leadership can ever be realized among blacks until the natural neurological disequilibrium position in black people is, as proposed earlier, scientifically rectified. Or plausibly, through random natural selection, Plato's ideal leader will emerge from the few intelligent blacks to cause dramatic change of mind-set to the rest of the black population.

The physical environment has been blamed for far too long for being the cause of black people's general insensitivities to public needs compared to nonblacks. The idea of blaming the environment should be seen as obsolete. The notion's advocates have not as yet attempted to cause a change in black people's mind-sets. Blacks have inherited the same mediocre comfort level from our ancestors and are ever ready to pass them on to future generations. Black people and their leaders at every level must firstly acknowledge that their mind-set changes are too long overdue.

The levels of progress among most blacks and the levels among most nonblacks are getting wider away from each other every day. Most nonblacks are wide awake and constantly probing unknown frontiers successfully, while most Negroes are deeply asleep and waiting to be knocked on the head to wake up and come to show participation in human progress, starting from the individual self. It is sadly true that most blacks have been anxiously waiting for God's intervention ever since European Jews succeeded in converting blacks into Christianity.

# CHAPTER 10

# Conclusion

I HOPE YOU, the reader, have carefully digested completely the content of this work. Judgment as to whether I successfully stated my point of view clearly and also made a convincing case regarding my observation of my own race of people would obviously depend on your nature and your intuition.

I am conscious of conventional wisdom that genuine observations scientific in nature of topics concerning natural organisms and human beings specifically need to be supported by well-researched or refined statistical data as evidence. Yes, this piece of work is full of scientific overtones. Yes, I am not a scientist by any means.

The work should have been done by an expert in human anatomy, a geneticist, a neurologist, a pathologist, a biochemist, or indeed, a specialist with a natural-science base.

I am aware also that some experts in the science field have done researches to indicate that natural similarities, equalities, and differences do exist among humans and other living organisms. But that which I have not come across yet is data showing a natural correlation between intelligence and the amount of melanin content in human beings, especially in human neurons and brain tissues and receptors to the brain organ for storage and processing of the received information and its transmission into human actions. There is no denial of their relationship, but their racial comparative quality and how related they are is unclear.

It is a fact that black people have more melanin pigment in the brain organ, in the neurons, in the skin, and in the hair than any race of people, especially more than Caucasians. What makes this excessive pigmentation of melanin in black people so significant is the fact that throughout human history, black people have shown no significant improvement in life. The brain power of most black people appears to be incapacitated by this element.

I have attempted to analyze melanin's relationship with performance and also reject the idea of the environment as the cause of black people's docility to white people. It is this substance which defines the nature of the Negroes. Our mental weaknesses, our lack of progress, our lack of foresight, our perpetual dependence on nonblacks, our inability to know what we have and how to exploit our own natural resources for our own benefit—all these and many more represent black people's nature.

E. ASAMOAH-YAW

And based on the natural instinct of most nonblacks, especially Caucasians, who mostly aspire to leave indelible legacy after death, such as someone whose name would be remembered by future generations as an originator, an inventor, a discoverer, a leader who changed people's mindset to become progressive, or a drug inventor or something in medical science to reconstitute Negroes' nature or modify an internal organ of the blacks into rational equilibrium with the rest of humanity.

The twenty-first-century world is so different that scientists understand human nature now better than before. Dedicated scientists are researching every day, looking for answers to solve human genetic problems. Hence, it is most probable that in the near future, the natural ailment which most black persons suffer from will be rectified to improve or indeed change completely through medical science. Black people must *first admit* that *there exists* a natural imperfect condition in *most* black people and allow science to probe for everlasting solution.

Nature's course can be changed. Many examples abound everywhere in several organic environments. Genetic engineering, stem cell manipulation, neurological stimulations, and several biochemical researchers exist today, which can be applied to change forever a black person's natural difference.

I am highly convinced that among all flying creatures, parrots are the most intelligent birds. I am also sure that among all aquatic animals, pink dolphins are the most intelligent. When it comes to human intelligence test, Negroes will definitely be at the bottom category. This conclusion is based on black people's negligible contribution to human progress and performance

E. ASAMOAH-YAW

history in comparison with those of nonblack people. And in the same order, as stated above with the birds and fishes tests, the best of the best among humans will not be a black person.

# BIBLIOGRAPHY

Hochschild, Adam, *King Leopold's Ghost* (Houghton Mifflin Co., 1998). ISBN #0395759242

Lewontin, R. C., Steven Rose, and L. J. Kamin, *Not in Our Genes: Biology, Ideology, and Human Nature* (Pantheon Books, 1984).

Darwin, Charles, *The Origin of Species* (Barnes & Noble: New York, 2004).

*Webster Ninth New Collegiate Dictionary* (Merriam-Webster, 1987).

O'Brien, P. K., *Encyclopedia of World History* (2000).

Campbell, Neil, Jane Reece, and Lawrence Mitchell, *Biology*, 5th edn (Jim Green: Menlo Park, 1999).

Middleton, Harris, and others, *The Black Book* (Random House, 1974).

Boahen, Albert Adu, *Topics in West African History*, Schools edn (Longham).

Brantlinger, Patrick, *Rule of Darkness (British Literature and Imperialism) 1830–1914*.

Radel, Stanley, and Marjorie Navidi, *Chemistry*, 2nd edn (West Publishing Co.).

Santrock, John, *Life-Span Development*, 13th edn.

*Peoples of the Earth Vol. 2: Africa from the Sahara to the Zambasi*.

E. ASAMOAH-YAW

# INDEX

birds   xiii-xv, 11, 13-19, 24,
      28, 193
*Black Book, The* (Harris,
      Smith, Levitt, Furman,
      Morrison)   xvi, 196
blacks   xvi-xvii, xix, 45-50,
      57-60, 67, 93-4, 114-15,
      120-1, 124-7, 140-1,
      143-4, 156-8, 169-72,
      183, 185-7
blacks:
   African   102, 167
   brain of   161
   nature of   viii, 52, 60, 65,
      95, 100, 190
brain   2, 14, 27-8, 30-1, 39,
      43-4, 53, 65, 72, 76,
      78, 111, 119-21, 161,
      189-90
brainpower   21
British   139, 141, 196
British prisons   128, 139

## C

Cape Mesurado   132
Casement, Roger   152, 155

Caucasians   53, 87, 93, 143,
      161, 167, 170, 174, 183,
      190-1
Chinese dynasties   88
Christianity   46, 168, 187
colonies   90, 141-2, 145-6,
      155, 169
communities:
   African   47, 62
   black   xvi-xvii, 57, 61,
      118, 127
   nonblack   38, 67, 109
Congo chiefs   149
Congo Free State   144-5,
      149, 154-6, 159
Conrad, Joseph   152, 155
Continental Congress   113
countries   35, 57, 108, 122,
      127-30, 136-9, 144,
      146-8, 162
   African   56, 64, 136, 162
   black   59, 165
   nonblack   58-9, 130, 184
creatures, aquatic   19-20
Cromwell, Oliver   89

fishes   11, 19-24, 28, 44, 86

Franklin, Benjamin   113

Freetown   132

Furman, Roger   xvi, 196

## G

generations   34-6, 42, 117, 173, 178, 180, 185, 187, 191

genes   1, 14, 17-18, 39, 72-3, 111-12, 116-17, 121, 157, 195

  lethal   73

  major groups of   72

  unique   14, 17

genetic engineering   10, 73, 76, 110, 192

genome, human   109, 158

genus   7, 11, 13, 74

Ghana   56, 62, 91, 162, 165, 170, 179, 182

God   45-6, 62, 65, 109, 114, 143, 187

Great Britain   138

## H

habitats   1, 4, 11, 14, 17-18, 21-2, 29, 33, 35, 38, 40, 43, 65, 86, 133-4

  natural   40, 133-4

Harris, Middleton   xvi

Harrison, William Henry   154

Henry the Navigator (prince of Portugal)   168-9, 171

Herrnstein, Richard   76

history   xvi, 79, 83, 87, 99, 114, 117, 125, 140, 156, 161, 167, 173, 179, 195-6

  human   173, 182, 190

homeland   133

hominid   81

*Homo erectus*   81-3

*Homo habilis*   81, 83

*Homo sapiens*   12, 24, 38-9, 74, 81, 83, 86, 91-2, 94, 96, 98, 102, 141

E. ASAMOAH-YAW

melanocytes 70-1, 75, 111-12

Melbourne 139

Mongolians 88

monkeys 11-12, 29, 42, 83

Monroe, James 132

Monrovia 132

Morel, Edmund Dene 152, 155

Morrison, Toni xvi, 196

Moses (Hebrew patriarch) 131

Murray, Charles 76

# N

natural order 25, 29

natural performances 112, 165

natural qualities 5, 13

natural sciences 107, 109

natural selection 116, 186

natural traits xviii, 26, 65

nature viii-ix, xvii-xviii, 2, 21-4, 26-7, 31-5, 37-40, 50-2, 54-5, 57-60, 67, 100, 105-9, 120-4, 190-2

biological 22-3

Neanderthals 81, 96

Negroes 23, 54, 58, 62-3, 83, 95, 103-4, 111, 113-14, 118, 122, 174, 181-2, 187, 190-2

nerves 15, 30, 119-20

nests 12-13, 17-18

neuromelanin 72, 112

neurons 14, 23, 38, 51-2, 76, 102, 118-19, 121, 143, 158, 165-6, 175, 183, 189

Niger River 155, 179

Nkrumah, Kwame 165

nonblacks:

Australian 163

nature of 176

Northern Ireland 138

nurture xviii, 2, 10, 27, 32, 36-8, 106, 117, 185

## O

organisms   5-9, 40, 73-4, 189

organogenesis   123

## P

parrots   13, 19, 192

people, nonblack   xvi-xvii, 23, 48, 57, 59, 65-6, 93, 101, 107-8, 110, 129, 137, 159, 173, 193

pheomelanin   71, 112

planet earth   25, 28, 40-1, 80-4, 137

Plato (Greek philosopher)   178-9, 186

populations   63, 71, 90, 130, 138

  black   63, 130

  human   35, 42, 113

Portugal   155, 167

Portuguese   89, 132, 167-9, 171-2

prisoners   128, 139, 152

## R

races   16, 25, 27, 49, 55, 66, 74, 104-5, 163, 170, 188, 190

racial groups   19, 26, 39, 76, 85, 87, 104, 125, 176

racial heritage   77

racial inequality   77

rivers   45, 61-2, 98, 100, 146, 168

Roberts, Joseph   133

Roman Empire   89

## S

Sahara Desert   106, 179

Scotland   138

Second World   108

SED (similarities, equalities, and differences)   5, 7-8, 11-13, 16, 19-20, 23, 29, 32, 36, 93, 104, 189

settlers   35, 49, 83, 101-2, 139-40, 161

  black European   49

  European   101-2

www.ingramcontent.com/pod-product-compliance
Lightning Source LLC
Chambersburg PA
CBHW031849200326
41597CB00012B/335